CLIMATE CHANGE
CHANGE

THE FACTS 2025

CLIMATE CHANGE

THE FACTS 2025

EDITED BY
JOHN ABBOT &
JENNIFER MAROHASY

CONTRIBUTORS

John Abbot   Michael Asten   Petr Chylek   Arthur Day
Manvendra K. Dubey   Hermann Harde   Aynsley Kellow
William Kininmonth   James D. Klett   David R. Legates
Glen Lesins   Jennifer Marohasy   Christopher Monckton
Antero Ollila   Ian Plimer   Tom Quirk   Jan-Erik Solheim

Institute of Public Affairs

First published 2024

Institute of Public Affairs
Level 2, 410 Collins Street
Melbourne, Victoria 3000
Phone: 03 9600 4744
www.ipa.org.au

National & International Distribution
by Australian Scholarly Publishing
scholarly.info

ISBN: HB 978-1-923267-08-4
ISBN: PB 978-1-923267-09-1

Cover photograph: Sunset in the Coral Sea on 10 June 2021 by Jennifer Marohasy
Cover design by Dimitri Mathieu, typesetting by Midland Typesetters and
printing by Everbest Printers

Contents

Foreword

Scott Hargreaves

The book you have in your hands is the distillation of 15 years of continuous research by the Institute of Public Affairs (IPA) on the fundamental question: how anthropogenic is anthropogenic global warming (AGW)? That is, how much influence do human activities actually have on the climate of the Earth?

It is the fifth in the series of books entitled *Climate Change: The Facts*, and the IPA is grateful to all the editors and contributors who have made them such vital publications, and to the ordinary men and women across Australia and the world whose financial support made publication possible. We are proud to record the names of those who supported this edition through donations on pages 235–236, and on behalf of the IPA I express my deepest appreciation to them.

I say the *Climate Change: The Facts* series is vital because climate policy remains the single biggest policy obsession of international institutions like the UN, the transnational organisations that shape public opinion in the West, and the leadership of the majority if not quite the totality of leadership of the various countries of the West. This in turn is having systemic impacts on trade flows, the geopolitical balance, economic growth, and inequalities of wealth and power within society.

The way we *frame* the fundamental question is also vital. We may be wont to say 'but the climate is always changing', which is a true statement, but it doesn't go far enough. As proponents of the AGW hypothesis have said, observing that there have been changes in the past does not necessarily refute the possible effects of recent increases in the concentration of CO_2, which is in turn attributed to industrialisation and agriculture.

Generally, scientists in the AGW camp have the integrity to acknowledge the variations, while sticking to their guns, including the inevitable quotations of what the models say *will* happen. To take an example, the abstract of Marcott et al. (2013) states:

> Surface temperature reconstructions of the past 1500 years suggest that recent warming is unprecedented in that time. Here we provide a broader perspective by reconstructing regional and global temperature anomalies for the past 11,300 years from 73 globally distributed records. Early Holocene (10,000 to 5000 years ago) warmth is followed by ~0.7°C cooling through the middle to late Holocene (<5000 years ago), culminating in the coolest temperatures of the Holocene during the Little Ice Age, about 200 years ago. This cooling is largely associated with ~2°C change in the North Atlantic. Current global temperatures of the past decade have not yet exceeded peak interglacial values but are warmer than during ~75% of the Holocene temperature history. **Intergovernmental Panel on Climate Change model projections for 2100 exceed the full distribution of Holocene temperature under all plausible greenhouse gas emission scenarios (emphasis added).**

The research in this book provides a much more complete answer to the question of how awareness of natural variation influences our view of the AGW hypothesis. Firstly, it looks at the question over a variety of timescales. Secondly, it treats the natural variations as *data* to be interrogated. It is not enough, as the paper just quoted appears to do, to passively observe that there have been variations in the pre-industrial past. The scientific method is to inquire: 'what might be the cause of those variations?' Notwithstanding those who denigrate such questions, it is legitimate to look at matters such as the behaviour of the Sun, the effects of planetary bodies, volcanism, and indeed whatever else might be relevant. Thirdly, based on the answers to that second question, we can begin to generate counterfactuals – alternative 2025 climates imagined without the contribution of industrialisation – against which we can assess the contribution of 'greenhouse gases'.

It occurred to me when re-reading the contribution of Dr Nicola Scaffetta in an earlier volume, *Climate Change The Facts 2017*, that the average consumer of climate news lives in a Copernican solar system. It is sometimes forgotten that while Nicolaus Copernicus (1473–1543) established the heliocentric view and overturned the Ptolemaic system in which the Earth was at the centre, his system had the planets revolving around the sun in regular (circular) orbits. And so it is that humans may look to but not beyond the sky, assuming a stability in the heavens which is not there, and never has been. Only a few of the planets are visible, and then only at night. They are small to the eye, and outside of consciousness when it comes to considering the matters that affect the Earth. Even the Moon and the Sun, while much more prominent in the sky, somehow escape consideration on most occasions when matters of climate change are officially discussed.

Copernicus was seeking a certain sort of order for his system, and the search for order and regularity is perhaps a feature of the human mind. Whereas what we now *know* is that orbits are ellipses, with a degree of irregularity, and the interaction of planetary bodies produces perturbations (quite apart from the dynamism and variability of the Sun itself). Why shouldn't we consider what this means for the climate?

I'm describing the focus of this book. This focus was developed after a series of conversations between the Editors, Dr Jennifer Marohasy, Dr John Abbot and myself. In its early formulations we had envisaged something more expansive, with essays supplementing the scientific investigations. These proposed essays were concerned with the psychological or sociological roots of climate catastrophism and its pernicious effects in other areas of public policy.

The book evolved however, to have a much tighter focus on the historical record of climate change, and how we might draw upon that to shed light on the AGW hypothesis. We decided it was most important to produce a book with a coherent theme and consistency in subject matter.

Therefore, even my own planned contribution, which was to have been on the topic of climate hysteria as an outgrowth of the 'de-growth' movement – as demonstrated by much of Greta Thunberg's *The Climate Book* (2022) and other publications – was excluded. I was keen to analyse the premises of *Degrowth can work – here's how science can help*, which appeared in *Nature Magazine* in December 2022 (v612). I planned to ask: what was an article premised on a Malthusian and erroneous conception of growth doing in what purports to be a journal of science? Might this indicate that other sociological and psychological factors are at work?

Two good things happened as a result of this tightening of focus. First of all, we decided that the more expansive contributions we had received, which explored the connections between human factors, science, climate and energy policy making, would be moved into a separate volume to be published by the IPA in early 2025. It is being edited by myself and IPA Senior Fellow, Dr Kevin You, and includes chapters from Dr Peter Ridd, and the IPA's Visiting Fellow in Energy Security, Stephen Wilson. The working title is *After the Failed Consensus: Rethinking Climate and Energy Policy for 2050*.

The second effect was that it enabled Dr John Abbot to recruit even more authors for the volume whose research is aligned with the more focused editorial approach. All along Dr Abbot had been adamant there were scientists toiling away in the universities, research institutions, or on their own dime, reaching broadly similar conclusions, and that their articles continued – somehow – to be published in academic literature. Firstly, he undertook an extensive literature search, identified key articles and approached the authors. Once the purposes of the book were explained, they were, in most cases, happy to contribute. In the period since the original notices for this publication were issued, we have thus been able to add as authors Dr Michael Asten, Dr Arthur Day, Dr Hermann Harde, Emeritus Professor Aynsley Kellow, and Dr Tom Quirk. We also welcomed on board James D. Klett, Glen Lesins, and Manvendra K. Dubey, who co-authored the chapter with Petr Chylek.

The longest feasible timescale on which we can consider the Earth's climate is, of course, the age of the planet on which we all live. For this reason, I am especially grateful to Professor Ian Plimer for his contribution of Chapter 4, *The Geology of Climate Change*. Even those readers familiar with his work will appreciate this erudite and stylish summation of matters geological. Those who are new to Ian's remarkable contribution to climate realism will never be the same again. This is the detached and appropriate perspective from which to consider the current climate, and climate change:

> Planet Earth is in the Goldilocks zone where liquid water has existed for billions of years and, unlike other planets, does not have diurnal temperature extremes, thereby making Earth habitable. Notwithstanding, variation in temperature over the course of human habitation has been from icehouse to hothouse conditions, no different from modern latitudinal climate variation.

While there is coherence to the subject matter in this volume, the tone does vary. Professor Plimer is – famously –direct and to the point. Dr John Abbot, one of our esteemed editors, is cautious and careful to let the readers draw their own conclusions. He observes that the evidence he has gathered *may* have implications for the various policy recommendations put forward by the IPCC, and indeed they do! Dr Jennifer Marohasy, now editing her third volume in the *Climate Change: The Facts* series, brings her characteristic knowledge of and passion for the *practice* of science to bear in her contributions.

As publisher I would like to thank Drs Marohasy and Abbot for all of their work in the IPA's climate science research program, and in particular for their heroic efforts in pulling this volume together. It will make a lasting contribution to scientific literature and provide a much-needed corrective to narratives which treat weather as climate change, or climate change as inherently unusual or attribute it *a priori* to anthropogenic sources.

I trust you will find great value in this work and recommend it to friends and colleagues in your networks. You can direct them to our main website for the series, where they can obtain further information on where to obtain hard copies and online editions.

www.climatechangethefacts.com.au

Thank you for supporting the research of the IPA, either directly through donations, or indirectly by purchasing this book. You can learn more about the work of the IPA at:

www.ipa.org.au

Scott Hargreaves
Executive Director
Institute of Public Affairs
Melbourne, August 2024.

Contributors

John Abbot

John is a Senior Fellow at the Institute of Public Affairs with more than 120 peer-reviewed publications in international journals, including more than a dozen recent publications in climate science. John is interested in understanding how natural systems change, and the application of artificial intelligence, particularly using neural networks, for forecasting rainfall, and for separation of natural and anthropogenic temperature change. He is a graduate in chemistry from Imperial College, London (1973) with an MSc from the University of British Columbia (1978) and a PhD from McGill University (Montreal) (1982), and also a Master of Biotechnology degree from the University of Queensland. He also has law degrees, a Juris Doctor (2003) and Master of Laws from the University of Queensland (2004), and was admitted as a solicitor in 2004. He has a Master of Technology Management from Deakin University (1997) and has held faculty positions at Queen's University, Canada, the University of Tasmania, Griffith University, Queensland University of Technology and Central Queensland University.

Michael Asten

Michael received a PhD in geophysics from Macquarie University in 1977, and subsequently worked in academia and the mineral exploration industry in Australia and overseas. He was a Professor (Research), part-time, in the School of Earth Atmosphere and Environment, Monash University, Melbourne, Australia, retiring in 2015, and continuing as an

Adjunct Senior Fellow until 2021. He is now Research Director for Earth Insight, in Melbourne.

Over the past decade he has applied methods of time-series analysis to the study of natural cycles of climate change. In 2020 he was an Expert Reviewer for the Intergovernmental Panel on Climate Change Second Order Draft of the 6th Assessment Report (area of expertise natural cycles of climate change and climate sensitivity). Michael has authored or co-authored over 200 scientific papers. He has published on climate sensitivity as deduced from deep-ocean records of the Eocene period and is part of an international group of scientists investigating centennial and millennial natural cycles of climate change and the hypothesis that they are in part subject to astronomical control via cosmic ray flux on the Earth's atmosphere.

Petr Chylek

Petr is a researcher for Space and Remote Sensing Sciences at Los Alamos National Laboratory. He received his diploma in theoretical physics from Charles University, Czech Republic, and received his PhD in physics from the University of California in 1970. Prior to becoming a government researcher in 2001, Petr was a professor at several United States and Canadian universities, including the State University of New York, Purdue University, University of Oklahoma and Dalhousie University, Canada.

Petr has published over 100 first authored scientific papers on remote sensing, atmospheric radiation, climate change, cloud and aerosol physics, applied laser physics and ice core analysis. His work has been cited more than 4000 times. He is best known for his work in remote sensing, aerosols and climate change.

In 2006, Petr served as Chairman, Scientific Program Committee for The Second International Conference on Global Warming and the Next Ice Age held at Los Alamos National Laboratory. The papers presented at the 2006 Conference were published in a special section of the Journal of Geophysical Research – Atmospheres in 2007. Petr and co-authors presented a paper in 2007 at the meeting of the American Geophysical Union estimating climate sensitivity to doubled atmospheric carbon dioxide to be significantly less than the Intergovernmental Panel on Climate Change estimate.

(Robert) Arthur Day

Arthur is an earth scientist with a PhD from Monash University, Australia, in volcanology. He worked at the Australian Nuclear Science and Technology Organisation (ANSTO) developing methods for the safe immobilisation of nuclear wastes in collaboration with other scientists around the world. Arthur wants to see government policies based on evidence and science, untainted by partisan political or ideological agendas. He currently has no professional affiliations, and no career-related vested interests.

Hermann Harde

Hermann undertook studies with a focus on atomic and laser physics at the Technical University Hannover from 1966-1970 completing a Diploma in Physics. From 1971-1974 he was scientific assistant and obtained a PhD at University of Kaiserslautern. Hermann was Professor of Laser-Engineering at Helmut-Schmidt-University Hamburg from 1975 to 1981. He was Professor for Laser-Engineering and Materials Science at this university from 1982 to 2009. He was also Dean of School of Electrical Engineering, from 1995 to 1997 and Vice-President between 1997 and 1999, and held the Chair of Experimental Physics & Materials Science from 2001 to 2009. He was a Visiting Scientist at IBM Watson-Research Centre, New York, in 1990 and 1992, and a Visiting Scientist at Oklahoma State University, Oklahoma in 1995 and 1996.

Hermann retired from his university position in 2010 and has since concentrated on research in climate science, with more than 30 publications and reports. Topics include the influence of the sun on temperature trends, factors controlling atmospheric carbon dioxide levels and experimental investigations of greenhouse effects.

Scott Hargreaves

Scott is Executive Director of the IPA, appointed in 2022. Prior to joining the staff of the IPA in 2015, he worked in various public and private sector roles with a heavy emphasis on public policy development with a skew towards the energy sector, including with agencies of the Victorian Government, Meridian Energy, and Anglo American plc, and also

Origin Energy (where he was for a time a manager of sustainability). He contributed the final chapter to Climate Change: The Facts 2020. He has a Bachelor of Arts in Politics and Economics, a Post Graduate Diploma in Public Policy, an MBA from the Melbourne Business School, and a Master of Commercial Law.

Aynsley Kellow

Aynsley holds a PhD in political studies from the University of Otago, New Zealand. He is an Emeritus Professor at the University of Tasmania, where he taught from 1981–1984 and 1999 –2018. He was foundation Professor of Social Sciences in the Australian School of Environmental Studies at Griffith University from 1992–1998, a member of the Joint Academies Committee on Sustainability, and an expert reviewer for the United Nation's Intergovernmental Panel on Climate Change's Fourth Assessment Report. His books include *Transforming Power and International Toxic Risk Management*, and *Science and Public Policy: The Virtuous Corruption of Virtual Environmental Science and Negotiating Climate Change*.

William Kininmonth

William has a career in meteorological science and policy spanning more than 40 years. For more than a decade (1986-1998) he headed Australia's National Climate Centre with responsibilities for monitoring Australia's changing climate and advising the Australian government on the extent and severity of climate extremes, including the recurring drought episodes of the 1990s.

He has extensive knowledge of global climatology, the climate system and the impacts of climate extremes developed through more than two decades associated with the World Meteorological Organization (WMO). He was Australia's delegate to the WMO Commission for Climatology and more recently has been a consultant for implementation of its programs. He coordinated the scientific and technical review for the United Nations Task Force on El Nino following the disastrous 1997-1998 event and has participated in WMO expert working groups.

He was a member of Australia's delegations to the Second World Climate Conference (1990) and the subsequent intergovernmental negotiations for the Framework Convention on Climate Change (1991-1992). He had a close association with the early developments of the climate change debate. His suspicions that the science and predictions of anthropogenic global warming had extended beyond sound theory and evidence were crystallised following the release of the 2001 Third Assessment Report of the IPCC.

David R. Legates

David received BA, MSc and PhD degrees from the University of Delaware. He is a climatologist who specialises in precipitation and climate change as well as spatial statistics. Upon receiving his PhD, he became a Professor at the University of Oklahoma. He later moved to Louisiana State University and subsequently returned to the University of Delaware. Recently, he was on leave to the federal government where he served as the Assistant Deputy Secretary of Commerce for Environmental Observation and Prediction and was detailed to the White House Office of Science and Technology Policy as the Executive Director of the United States Global Change Research Program. He has been invited to speak to the US Senate Committee on the Environment and Public Works on three separate occasions. He also participated in the historic joint USA-USSR protocol for the exchange of climate information. He is now Emeritus Professor at the University of Delaware and works for the Cornwall Alliance for the Stewardship of Creation.

Jennifer Marohasy

Jennifer is a Senior Fellow at the Institute of Public Affairs, with publications in the international climate science journals *Atmospheric Research* and *Advances in Atmospheric Research*, as well as *Wetlands Ecology and Management, Human and Ecological Risk Assessment, Public Law Review* and *Environmental Law and Management*. Jennifer has a BSc and PhD from the University of Queensland (Brisbane) and a life-long interest in natural history and long-range weather forecasting. Jennifer blogs at jennifermarohasy.com

Christopher Monckton

Christopher Monckton has held positions with the British press and in government, as a press officer at the Conservative Central Office, and as Prime Minister Margaret Thatcher's policy advisor. He is a policy advisor to the Heartland Institute.

Christopher has given speeches, lectures, and university seminars on climate change around the world. He has testified four times before the U.S. Congress. He spoke at United Nations conferences in Bali, Bonn, Copenhagen, Cancun, Durban, Rio, and Qatar and was Nerenberg Lecturer in Mathematics at the University of Western Ontario in 2013. For his work on the climate he has been presented with numerous honours, including the Meese-Noble Award for Freedom, the Valiant-for-Truth Award of the Committee for a Constructive Tomorrow, the Santhigiri Ashram Award, and the Intelligence Medal of the Army of Colombia.

Christopher has authored papers on climate issues for the layman, as well as for peer-reviewed scientific journals. He established in a paper for the World Federation of Scientists that carbon dioxide has a social benefit, not a social cost. He was also a co-author of the paper that showed the claim of "97% scientific consensus" about climate change to be highly exaggerated. His latest paper exposes a substantial error in the computer models of climate that has led to wild official exaggerations of the high-end estimate of future manmade global warming.

Antero Ollila

Antero has an MSc in Process Engineering (1972) and a Licentiate of Science in Technology in Process Dynamics, 1974, at Oulu University, and Doctor of Science Technology in Quality Management, 1996 at Helsinki University of Technology. He has worked in various positions in industrial companies including Research and Development director and quality director positions. His latest work was as Adjunct Assistant Professor of Quality Management at Helsinki University of Technology.

He has been an active climate researcher since 2011 and published articles on climate change. He is a member of the Scientific Council of the Norwegian Association of Climate Realists. He has applied spectral analysis in greenhouse gas impact studies, and his main research subjects have been the greenhouse effect, the warming impacts of greenhouse gases, the carbon cycle, the energy balance of the Earth, and dynamical simulations of climate.

Ian Plimer

Ian was Professor and Head of Geology at the University of Newcastle (1985-1990), Professor and Head of Earth Sciences at the University of Melbourne (1991-2005), DFG (German Research Foundation) Professor at the Ludwig Maximilians Universität (München) (1991) and Professor of Mining Geology at the University of Adelaide (2006-2011). He is Emeritus Professor at the University of Melbourne. Ian held early career academic positions at Macquarie, New South Wales and New England Universities and, during mid-career and post academia, held senior positions and directorships in the exploration and mining industries.

Tom Quirk

Tom trained as a nuclear physicist at the University of Melbourne, Australia, attended the Harvard Business School, and has been a Fellow of three Oxford Colleges. During a long professional career, Tom at various times worked for resources company CRA (now known as Rio Tinto), and in the United States at Fermilab, and at the University of Chicago and Harvard, and at CERN (the European Organization for Nuclear Research). He has held several positions in utilities associated with electricity generation, including a founding directorship of the Victorian Power Exchange.

Jan-Erik Solheim

Jan-Erik is Emeritus Professor at the Arctic University of Norway, Tromsø, Norway. He has a degree in astrophysics from the University of Oslo (1964) and spent four years as researcher at McDonald Observatory and University of Texas. He was responsible for planning, building and running Skibotn Observatory, Norway and participated in planning the Nordic Opical Telescope at La Palma in Spain. His main research interests are cosmology, photometry of galaxies, astero-seismology of white dwarf stars and interacting binary stars; and recently Sun-Earth relations. He was chief editor for the international scientific journal *Science of Climate Change* 2022-23. He is a Member of Royal Astronomical Society and the Lithuanian Academy of Science.

Introduction

Dr John Abbot

Whenever issues of climate change are being presented, the media invaria-
bly delivers to the public the unquestioning perspective that any observed
changes are inevitably the result of human (anthropogenic) activities,
particularly through emissions of greenhouse gases. Discussion of the
possible contribution from natural factors to climate change is generally
absent. This is a view strongly endorsed by the Intergovernmental Panel
on Climate Change (IPCC); in this context one often hears the phrase
'the science is settled' and that the emphasis now needs to be on reducing
greenhouse gas emissions without further questioning. However, it
remains important to examine the changes in climate that have occurred
in recent times, particularly during the industrial era, in the context of
what is known about changes in the Earth's climate over much longer
periods. There are hundreds of studies published in the scientific literature that focus on how the Earth's climate has changed over hundreds
and thousands of years.

   Prior to the industrial era, extending back to about 1880 AD, the
Earth's climate underwent change on many different time-scales and
it is reasonable to assume that these changes were the result of natural
phenomena, examples of which can include the influence of the Sun and
the impact of volcanoes.

   It is also reasonable to assume that the Earth's climate would continue
to be influenced by these natural phenomena during the industrial era,

but that there may also be additional contributions from anthropogenic activities.

A key question that needs answering is to accurately evaluate the relative contributions from natural and from anthropogenic factors. However, it is important to appreciate that the Earth's climate system is highly complex, and consequently a number of different factors need to be examined when trying to answer this question. The IPCC's perspective – that human emissions of carbon dioxide (CO_2) are essentially acting as a 'control knob' on temperatures – is a very simplistic scientific scenario, but one that is easily presented to the public.

Proxy temperature records and cycles in climate

Studies that generate records of past climate, particularly of temperatures, have been produced for different parts of the world for the Holocene – the period spanning the last 10,000 years. These are derived from measurements of preserved physical characteristics of the natural environment – including, for example, tree rings, ice cores, stalagmites and stalactites found in caves, corals, marine sediments and lake sediments – known as proxy records. Proxy records are very important as they extend far beyond the temperature records that we have from instrumental thermometers, which at best only span the past 200 years and then (for that 200-year period) only in very few locations worldwide.

An important characteristic seen in many proxy temperature records is the presence of repeating cycles (or oscillations) on different timescales, and these are embedded within the resultant temperature profiles. For example, considering European proxy records over the past 2000 years, oscillations with maxima corresponding to the Medieval Warm Period and Roman Warm Period are clearly evident. In the first chapters of this book, we look at these cycles in more depth. In addition to the proxy records, these warm periods were also recorded directly in the historic lived experiences of people at the time. For example, the Romans tending vineyards as far north as Hadrian's Wall, and the Vikings growing crops in Greenland where it is no longer possible today, and where their graves are now in permafrost.

In chapter 1, I introduce cycles over 2000 years. In chapter 2, retired Professor Jan-Erik Solheim, Institute of Theoretical Astrophysics, University of Tromsø, Norway, discusses climate cycles of similar length that can be determined from the ice edge in the Barents Sea. I illustrate how the identification of natural cycles within proxy records can assist in the separation of natural and anthropogenic components of climate change in chapter 3. Several chapters provide support for the idea that the natural cycles are related to changes in solar activity (chapters 1, 2, 4). In chapter 4, Ian Plimer, Emeritus Professor of Earth Sciences at the University of Melbourne, Australia, extends the discussion to include natural cycles over millennia.

Limitations of general circulation models

Models – in which it is assumed that there is good understanding of physical processes involved in the Earth's climatic system and that these can be expressed mathematically – have played a key role in describing climate change as supported by the IPCC. These models, known as general circulation models (GCMs), involve complex computational processes and are generally run on supercomputers. Output from these models has supported the contention that climate change during the industrial era is predominantly anthropogenic and caused by human emissions, primarily CO_2. However, the IPCC's reliance on GCMs is questionable.

In chapter 5, Christopher Monckton, the third Viscount Monckton of Brenchley, examines why climate models fail, and emphasises the extent to which the assumed present climate emergency is based entirely on GCMs – a fact that is not widely appreciated. Monckton draws attention to two major GCM defects. The first is the failure to account for the propagation of uncertainty, while the second is a tenfold overstatement of feedback strength.

In chapter 6, Aynsley Kellow, Emeritus Professor School of Social Sciences, University of Tasmania, Australia, considers evidence that the climate models exaggerate temperature increases because they incorporate assumptions of equilibrium climate sensitivity (increases in mean global temperatures for a doubling of atmospheric CO_2) that are too high, largely because of the assumptions about positive feedback from water vapour, for which there is a lack of evidence. Another limitation of GCMs is that they do not represent the cyclic temperature profiles that exist in proxy temperature records.

In chapter 7, Professor David Legates, University of Delaware, USA, discusses the limitations of GCMs in representing precipitation, which includes rain, sleet, snow, ice pellets and hail. Given the importance of air temperature, it is justifiably a major focus when discussing climate change trends and predictions. But why isn't precipitation also presented? Usually, it is simply stated that a warming world is a more variable world regarding precipitation, with more floods and droughts, but maps and graphs of precipitation patterns and trends are rarely portrayed on websites and in the media. Four types of precipitation-forming processes exist, each requiring the air to cool below the water vapour saturation temperature. One of the many problems associated with precipitation simulation within GCMs is that only one of these four mechanisms – surface convection – produces virtually all of the precipitation. Precipitation simulated by a GCM occurs relatively frequently over large regions and producing rainfall that occurs too often with an intensity that is too low. Another fundamental problem of GCMs is that they exhibit considerable intra-model variability, but lack inter-annual variability. Simply put, the models vary considerably among themselves, but each year within the model looks very much like every other year.

My chapter 3 shows artificial intelligence (AI) forecasts are more skilful than climatology (long-term average) for rainfall, while the POAMA (Predictive Ocean Atmosphere Model for Australia) GCM that ran on very expensive supercomputers was on average less skilful than simply calculating a long-term average for each month. An example of a successful application of AI to weather forecasting is the prediction of monthly rainfall at long lead times associated with the flooding of south-east Queensland. Considering the town of Miles in Queensland as an example, there is a good correspondence between artificial neural network (ANN) forecasts at twelve months lead time and the actual monthly rainfall.

Carbon dioxide in the Earth's atmosphere

The level of CO_2 in the Earth's atmosphere has increased during the industrial era. However, the assumptions that this is predominantly through anthropogenic emissions, which is thought to be the predominant cause of any rise in temperatures, needs to be carefully examined. In chapter 8, retired meteorologist William Kininmonth discusses the characteristics of recent climate change. The systematic collection of satellite observations made over the most recent four decades has provided a powerful database for analysing the changing climate. The regional and seasonal characteristics strongly suggest that natural processes have been a primary cause for recent warming. The global near-surface temperature trend is 1.7 °C/century, but this value masks significant zonal differences. The warming for the equatorial band is only 1.1 °C/century, whereas the trend for the Arctic is 3.1 °C/century and that of the Antarctic is 4.1 °C/century. The regional trends are significantly different from the global average. These differences are masked by using the metric of global average temperature anomaly as an index of climate change. Any explanation for recent warming must acknowledge and accommodate these regional differences.

The role of greenhouse gases in the atmosphere (water vapour and CO_2, plus other minor gases) is often presented as absorbing radiation emitted from the surface to keep the atmosphere (and Earth) warm. This is questionable. Greenhouse gases do absorb radiation emitted from the surface, but they also emit radiation, both to space and back to the surface. The emissions to space and back to the surface exceed the absorption; that is, greenhouse gases tend to cool the atmosphere. The energy source of the observed atmospheric warming is an increased rate of flow of heat and latent heat from the tropical oceans to the atmosphere as the oceans have slowly warmed. Latent heat is heat energy stored in the transition from liquid water to water vapour, and released again upon condensation. The wintertime polar amplification of warming has occurred because of the increased rate of input of latent heat from the tropical oceans and enhancement of the natural seasonal cycle of poleward energy transport.

In chapter 9, Professor Hermann Harde, Helmut-Schmidt-University Hamburg, Germany, addresses the question: how much human or natural emissions are responsible for the observed increase in greenhouse gases (GHG), considering, in particular, the concentration of CO_2 in the atmosphere? Analysis shows that the anthropogenic contribution to increased CO_2 can be expected to add less than 10%, which is in strong contradiction to the IPCC's view.

The observed CO_2 evolution, inclusive of its annual cycle, has recently been reproduced in numerical simulations. This shows how the abundance of atmospheric CO_2 is controlled by a competition between two opposing influences – the feed of CO_2 through emission, and its removal through absorption at the Earth's surface. This competition governs time-mean CO_2, where absorption features centrally. It determines if and how fast CO_2 grows, as well as the magnitude of its perturbation, for example, by anthropogenic emissions.

The global temperature today is approximately ~0.7 °C warmer than it was half a century ago. Satellite records of global temperature indicate that, over much of the Earth's surface, temperature underwent no systematic heating during the last four decades. Perceptible heating was only introduced during just two brief intervals, both for not more than about two years in length: one preceding the El Niño in 1997, the other preceding the El Niño in 2016. It would be virtually impossible for such heating twenty years apart to be caused by continuously emitted anthropogenic GHGs, which are mainly released in the northern mid-latitudes.

In chapter 10, Adjunct Professor Antero Ollila at Aalto University School of Science and Technology, Finland considers the relative contributions to current global warming attributable to natural versus anthropogenic climate drivers. The prevailing IPCC theory is that anthropogenic global warming (AGW) is the dominant cause of warming, with greenhouse gases from human emissions being the major drivers. Evidence shows that the contributions to global warming from natural drivers are significantly underestimated by the IPCC. This scenario gives rise to an alternative theory called Natural Anthropogenic Global Warming (NAGW).

Century- and millennial-oscillations have a major role in explaining long-term variations during the anthropogenic period from 1750 AD to the present. The Earth receives about 99.97% of its energy from the Sun. The Sun's radiated energy measure is the total solar irradiance (TSI), which has both long-term variations in the millennium scale and short-term variations, such as the 11-year Schwabe sunspot cycle. Research studies show significantly lower radiative forcing (RF) and climate sensitivity values for anthropogenic climate drivers than assumed by the IPCC. Natural climate drivers have been suggested as the partial or total solution for global warming, including solar radiation changes, cosmic forces, and multi-decadal, and century- and millennial-scale climate oscillations. The cloudiness changes seem to have a significant role in magnifying cosmic effects like the TSI changes. According to the AR6 (IPCC 2021), the contribution of CO_2 during the industrial era has been 1.01 °C, but according to this study by Ollila it is 0.36 °C.

The CO_2 released into the atmosphere by plants and that released by the oceans into the atmosphere have distinctive isotopic signatures, as described in chapter 11 by retired Professor Michael Asten, School of Earth Atmosphere and Environment, Monash University, Melbourne and co-author physicist Dr Tom Quirk. Plants, for example, preferentially take up the lighter isotope of carbon (^{12}C) during photosynthesis. It is not possible to distinguish the plant component of CO_2 – as released, for example, by rotting plant material, phytoplankton productivity, and forest and peat fires – from that released by the burning of fossil fuels. However, it is possible to distinguish the ocean component that is often confounded with human emissions. The ocean and plant components, as derived from isotopic data, are approximately equal, at 50% of the annual CO_2 concentration increase.

In chapter 12, Dr Jennifer Marohasy, Senior Fellow, Institute of Public Affairs, Australia, discusses the idea that even a doubling beyond current concentrations will only cause about 0.71 °C increase in temperatures – a figure that is too small to matter. Dr Marohasy argues that what is not widely understood is that the 'catastrophe' in the IPCC models derives not from radiation transfer, but from assumed huge positive feedback that can be considered exaggerated. This analysis shows that even a doubling of CO_2 cannot cause a climate catastrophe, and that there are benefits for plant growth caused by increasing atmospheric concentrations of CO_2 – the greening of the Earth, which has now been directly observed by satellite.

Importance of volcanic eruptions

There is strong evidence for the existence of natural cycles or oscillations within the records of temperature that extend back thousands of years. These cycles extend into the industrial era and to the present day and continue to exert an influence on Earth's climate. Their combined effects can explain a significant proportion of warming over the past several decades through natural processes. There are other natural influences that may occur sporadically that can affect climate, such as volcanic eruptions.

As discussed by Dr Arthur Day in chapter 13, even some moderate eruptions can trigger rapid cooling. Ice cores containing a record of temperatures over time confirm that the coldest decades of the past 2500 years are associated with volcanic episodes. During an eruption, volcanic material injected into the stratosphere, the second layer in the atmosphere, can remain aloft for a very long time. 'Big' eruptions in recent decades are relatively small in comparison to those in past geological records. Cooling impacts of very big eruptions have been sufficient to cause continent-wide crop failures and widespread famines. This chapter examines several recent major volcanic eruptions and their influence on climate. These included the Hunga Tonga–Hunga Ha'apai eruption in January 2022 that had no measurable global temperature impact and the Mount Pinatubo eruption in June 1991 that did have a measurable temperature impact.

After major eruptions, winter warming in the latitudinal belt 50 to 70°N has been observed, as discussed in chapter 14 by Dr Petr Chylek, Earth and Environmental Sciences, Los Alamos National Laboratory, USA, with co-authors James Klett, Glen Lesins and Manvendra Dubey. Such warming was especially high over Eurasia for two winters following large volcanic eruptions in 1982 and 1991. Current state-of-the art climate simulation models reproduce the winter warmings after Mt Pinatubo. However, they do not reproduce the warming after the

El Chichon eruption. More than half of the individual climate models indicate cooling where warming is actually observed after the El Chichon volcanic eruption. This suggests caution in using any individual global climate model, or an ensemble of global models, for the study of specific regional climate phenomena, including any postulated regional effects of volcanic eruptions.

Application of artificial intelligence to weather and climate forecasting

AI has featured prominently in the media, with the growing availability of software and awareness of the wide diversity of potential applications. Machine learning (ML), a form of AI along with ANNs, has been used to forecast climatic variables including temperature and rainfall. Many studies using ML for climatic forecasts have been published in the scientific literature over the past decade. However uptake of these forecast methods by agencies such as the Australian Bureau of Meteorology has been slow, with an attendant reluctance to acknowledge the limitations of GCMs. These are complex physical models run on very expensive supercomputers that assume a good understanding of the underlying physical processes. In contrast, ANN methods rely on learning from historical data. ANNs enable patterns and relationships embedded within the historical data records to potentially be used to generate forecasts without a prior understanding of their physical origin. Physical models can pre-impose relationships and constraints that actually may not be valid and consequently restrict the capacity of the model to make accurate forecasts.

In chapter 3, I provide three examples of the application of AI to climate and weather forecasting. One successful application of AI to weather forecasting is the prediction of monthly rainfall at long lead times.

In another example AI and component oscillations were used to forecast natural temperature changes from proxy temperature records from the past 1000 years. The ANN was trained using data up to 1880 AD. Industrial-era forecasts were generated using the trained ANNs and extended oscillations as input data in order to forecast profiles from only the natural contributions to temperature change. The increase in temperature after 1880 AD was predominantly due to a continuation of the influences of natural factors, also present during the pre-industrial period. My chapter also reviews studies that show improved AI forecasting of cyclone trajectory and landfall location and also forecasting of cyclone intensity, winds, and storm surge.

Natural and anthropogenic climate change

As described by me in chapter 3, in the Northern Hemisphere and regional southern South American, proxy temperature records can be decomposed into sets of oscillations that can be used to forecast temperatures into the industrial era, based on oscillatory patterns up to 1880 AD. This approach has also been applied to other regions, including Switzerland, Tasmania and New Zealand. These temperature projections indicate warming due to natural climate cycles would be in the range 0.6 to 1 °C, while the difference between the model outputs and actual recorded values was at most 0.2 °C. This approach indicates that the dominant cause of atmospheric temperature increase during the industrial era is due to the continuation of natural temperature oscillations that have been present during the previous 1000 to 2000 years.

The possible scenario that recent global warming during the industrial era is mainly caused by a continuation of natural influences, with only a minor contribution from anthropogenic emissions, potentially has profound consequences for the growing demands for financial compensation associated with fossil fuel use in developed countries. The concept of climate reparations is a topic of intensifying global discussion. It has been suggested that wealthy, high-emitting countries should assist developing nations with the costs of decarbonisation. It is also suggested that high-emitting nations should compensate other countries grappling with the damages inflicted as a result of climate change, including the impacts of extreme weather events including rising sea levels. According to one recently published analysis (Fanning & Hickel 2023) in the journal *Nature Sustainability*, high-carbon countries owe at least $192 trillion to low-emitting nations as compensation for their greenhouse gas pollution.

A recent review by Professor Jacqueline Peel, Professor of Law at the University of Melbourne (Peel 2024) focuses on the internationally increasing use of courts to hold governments and companies to account for climate change. This cites a report from the United Nations Environment Program which found 127 climate lawsuits for Australia. The report states that:

> Climate change litigation provides civil society, individuals and others with one possible avenue to address inadequate responses by governments and the private sector to the climate crisis.

An example in the report is the case brought by fourteen Torres Strait Islanders, six of whom were children, to the UN Human Rights Committee. In 2022 the committee delivered a finding that the Australian government was violating its human rights obligations to Torres Strait Islanders through climate inaction. This was the first time that a country had been held responsible for its greenhouse gas emissions under international human rights law.

If it were to be accepted that only a small fraction of climate change was directly associated with human emissions, as shown by many of the chapter authors in this book, and the major cause is natural then that may profoundly affect the outcome of these cases.

1 Cycles in Very Long Temperature Records
Dr John Abbot

Most instrumental temperature records are limited to about the past 200 years for only a few European locations (for example, Jones 2001; Luterbacher et al. 2016; Parker et al. 1992). Various proxies have been used to extend temperature records back throughout the Holocene that spans the last 10,000 years, based on information from a range of different sources, including tree rings, ice cores, speleothems (stalagmites and stalactites found in caves), corals, marine sediments, lake sediments, and historical documents (Hernández et al. 2020). These proxy temperature records have been reported at local, regional, continental, hemispheric and global scales (for example, Trachsel et al. 2010; Ljungqvist 2010).

Proxy temperature records during the Holocene, and particularly for the past 2000 years, provide a very important context for the analysis and understanding of climate change during the industrial era.

There has been ongoing debate and controversy over the past two decades regarding the profiles of multi-proxy temperature reconstructions, particularly for the Northern Hemisphere. This is focused on two apparently discordant perspectives of temperature records for the past 1000–2000 years. The first perspective dates back at least to the Intergovernmental Panel on Climate Change (IPCC) First Assessment Report in 1990, which showed that temperature change, represented by a cycle exhibiting a distinct Medieval Warm Period (MWP) peak, a Little Ice Age (LIA) trough followed by eighteenth, nineteenth and twentieth century temperatures, increasing, but not exceeding those from the MWP. The second perspective began with the emergence of a 'hockey stick' proposed by Mann and co-workers in 1998. This was favoured in the IPCC Third Assessment Report, and subsequently a polarisation of viewpoints has emerged, conducted in scientific literature, the mainstream media and within political arenas.

The presence of cycles with oscillatory characteristics within proxy temperature reconstructions across a range of timescales, including millennial, centennial and decadal, has been reported in many investigations. The present study examines oscillations from spectral analysis applied to published multi-proxy temperature records. The analysis shows that each record can be represented by a set of four to eleven sine curves from spectral analysis that includes a dominant millennial oscillation, and several centennial and decadal oscillations. The apparent divergence into either a hockey stick or MWP–LIA cycles is derived from the phase alignment of the centennial and decadal oscillations with respect to the millennial oscillation. The maximum temperature of the dominant oscillation at around 1000 AD is increased by superimposing the centennial and decadal oscillations for MWP–LIA cycles, whereas it is reduced, and the profile flattened for the hockey stick. This could explain why current temperatures may exceed any in the past 1000 years with the hockey stick profile.

Evidence for the occurrence of a MWP can be found in many investigations. However, there is now also evidence that during this period there were sub-regions of the globe, for example within parts of the Mediterranean and parts of Africa, which simultaneously experienced cooling. This can be explained in terms of differences in the oscillatory processes affecting the climate. Considering the results of present studies regarding oscillations, it is possible to explain both hockey stick and MWP–LIA cycles having concurrent validity if there is a weighting towards records representing different geographical areas, corresponding to different phase relationships between a dominant millennial oscillation and associated centennial and decadal oscillations. Thus, it is possible for both types to co-exist. This still leaves open the possibility that the differences in methodology for reconstructions may also play a significant role in determining the appearance of the hockey stick temperature profile.

By applying spectral analysis to generate the input needed to train neural networks, the projection of pre-industrial temperature oscillatory patterns beyond 1880 AD shows that current temperatures can be largely explained on the basis of a continuation of natural oscillations. This is the case irrespective of whether hockey stick or MWP–LIA cycles are operative. This process could give rise to temperatures higher than the past 1000 years without any major contribution from anthropogenic influences.

Much emphasis has been placed on composite Northern Hemisphere proxy temperature reconstructions for the past 2000 years during the past two decades, when considering climate change. These temperature records have been central to the IPCC Assessment Reports since 1990. Development of proxy profiles has been, and continues to be, very controversial both in the scientific domain and within the wider community. Broadly viewed, there are two apparently divergent paradigms.

One perspective is represented by profiles that are clearly cyclical with maxima and minima corresponding to a distinct MWP peak and LIA trough in the past 1000 years, and also, if the temperature record is extended back 2000 years, a Roman Warm Period (RWP) and the Dark Age (DA). Figure 1.1 shows the Northern Hemisphere temperature profile that was presented in the IPCC First Assessment Report in 1990. It shows the temperature oscillating over the past 1000 years, with highest temperatures reached during the MWP in about 1200 AD, the lowest temperatures during the LIA in around 1600 AD, followed by increasing temperatures moving towards the current warm period (CWP). Temperatures for the twentieth century are lower than the maximum reached during the MWP.

Studies by Mann, Bradley and Hughes (1998 – referred to as MBH98; and 1999 – referred to as MBH99) and others, during the past two decades have generated an alternative representation of global and Northern Hemisphere temperatures that have been widely publicised and often colloquially referred to as 'the hockey stick'. These reconstructions exhibit a slow, long-term, downwards cooling trend (the ice hockey stick) to about 1900 AD, followed by relatively rapid warming in the twentieth century (the blade), with the instrumental

Figure 1.1 The MWP–LIA cycle

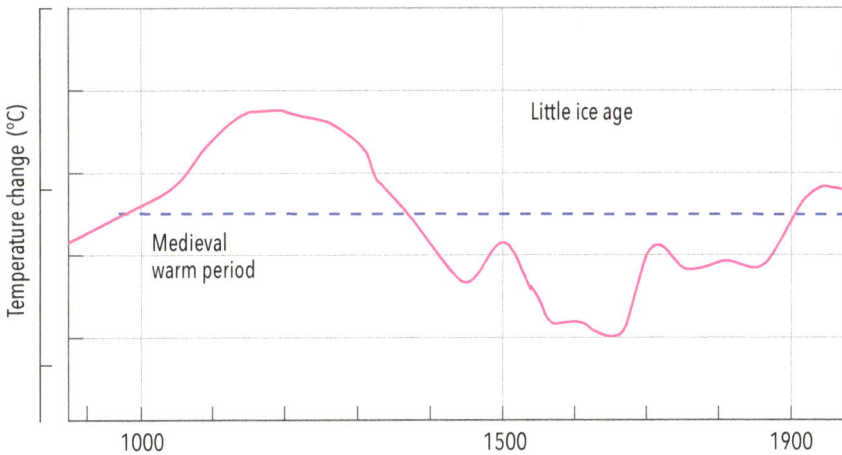

Source: IPCC First Assessment Report (Fig. 7.1c), Folland et al, 1990.

temperature record exceeding earlier temperatures during the previous 1000 years by 2000 AD. While the IPCC Third Assessment Report in 2001 (TAR) drew on five reconstructions to support its conclusion that recent Northern Hemisphere temperatures were the warmest in the past 1000 years, it gave particular prominence to an illustration shown in Figure 1.2 based on the MBH99 paper. The hockey stick graph was subsequently seen by mass media and the public as being central to the IPCC case for global warming and a developing climate crisis.

The IPCC Fifth Assessment Report 2013 (Ch. 5, Information from Paleoclimate Archives, Fig. 5.7) shows a set of temperature reconstructions that include both types, without drawing significant attention to the ongoing debate. The commentary states that the timing of warm and cold periods is mostly consistent across reconstructions (in some cases this is because they use similar proxy compilations), but the magnitude of the changes is clearly sensitive to the statistical method applied and to the target domain (land, or land and sea; the full hemisphere, or only the extra-tropics).

Nevertheless, the validity of the hockey stick graph has been intensely debated in the scientific literature over the past two decades (for example, Zorita et al. 2003; von Storch et al. 2004; Bürger et al. 2006; Zorita et al.

Figure 1.2 The 'hockey stick' graph

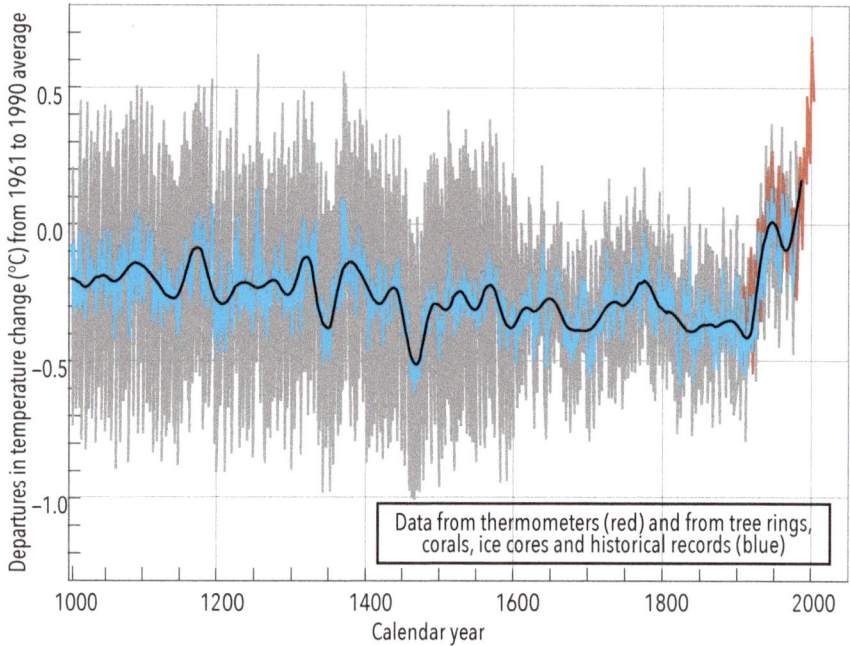

Source: IPCC Third Assessment Report (Fig.1) in Summary for policymakers 2001.

2007; Wahl & Ammann 2007). This debate includes methodologies used for the construction of multi-proxy temperature, the occurrence of the MWP and LIA, whether present temperatures are indeed unprecedented in the past 1000 years, and the causation of current warming trends.

The importance of the implications of the temperature reconstructions is also reflected in responses from the political sector. In the United States, the North Report (National Research Council 2006) evaluated reconstructions of the temperature record of the past two millennia, providing an overview of the state of the science and the implications for understanding of global warming. It was produced by a National Research Council (NRC) committee chaired by Gerald North, at the request of the U.S. House of Representatives Committee on Science. The NRC committee stated: 'The basic conclusion of Mann et al. (1998, 1999) was that the late twentieth century warmth in the northern hemisphere was unprecedented during at least the last 1000 years. This conclusion has subsequently been supported by an array of evidence that includes both additional large-scale surface temperature reconstructions and pronounced changes in a variety of local proxy indicators.'

Several studies have found evidence for oscillations in the proxy records at millennial, centennial and decadal timescales by applying spectral analysis (Ludeke & Weiss 2017; Abbot & Marohasy 2017a; Wilson et al. 2007). It has been suggested that some proxy records may not include low-frequency oscillations because certain types of proxy (particularly tree rings) do not capture this compared to other proxies, such as ice cores. The occurrence of oscillatory processes in past temperature records is important because it potentially provides a method of projecting forwards temperature patterns based on pre-industrial influences into the industrial era. The availability of this forecasting capability potentially enables the separation of natural and anthropogenic influences on temperature during the industrial era (Abbot & Marohasy 2017a). Identification of oscillatory patterns in temperature records are also useful in understanding potential linkages with solar activity that are known to exhibit oscillatory behaviour on similar timescales.

There is evidence that multi-decadal oscillations have an influence on recent temperature variations. For example, Wan et al. (2019) concluded that for Canada up to 0.5 °C of the observed warming trend may be associated with decadal variability of the climate, as represented by the Pacific Decadal Oscillation (PDO) and North Atlantic Oscillation (NAO). However, the presence and utility of oscillatory behaviour is a contentious issue. Mann et al. (2020) recently published an article claiming that such decadal and multi-decadal oscillations are merely 'noise' and questioned the validity of using oscillations in forecasting.

Hernández et al. (2020) recently reviewed the significance of decadal oscillations throughout the Holocene, including the El Niño–Southern Oscillation (ENSO), the Pacific Decadal Variability (PDV), the NAO, the Southern Annular Mode (SAM), the Atlantic Multi-decadal Variability (AMV), and the Indian Ocean Dipole (IOD). In citing Mann (2020), Hernández et al. (2020) report that 'the debate still rages' concerning decadal oscillations.

A focus of many proxy temperature reconstructions has been whether temperatures reached in the industrial era are unprecedented when compared to temperatures during the Holocene, particularly over the past 2000 years. For example, Ge et al.'s (2013) temperature reconstruction over a period of 2000 years for China found that the warming during the twentieth century is not unprecedented; similar warming occurred in 981–1100 AD and again in 1201–1270 AD. A study of the Arctic (van der Bilt 2019) spanning 12,000 years found early Holocene temperature oscillations exceeded the amplitude of the current observed and projected warming in Svalbard lakes, with temperatures up to 7 °C higher than they are today.

A number of studies have expressed concerns that the reliability of the conclusions drawn from the General Circulation Models (GCMs), which are relied on by the IPCC, are questionable because of apparent failures of these models to simulate the oscillatory patterns that are clearly present in both recent and paleo temperature records. Scafetta (2013) reported that some very important physical mechanisms necessary for reproducing multiple climatic oscillations, which are responsible for about half of the 1850–2010 AD warming, appear to be missing in simulations generated by GCMs. In a later investigation, Scafetta (2019) concluded that GCMs fail to properly reconstruct the natural variability of the climate throughout the entire Holocene. This shortcoming was evident at multiple timescales, including the large millennial oscillations observed throughout the Holocene that were responsible for the MWP and the shorter climatic decadal oscillations with periods of 10–60 years. De Larminat (2016) examined the limitations of GCMs and questioned IPCC's conclusions because the studies did not look beyond the industrial era, concluding that they were inadequate regarding the influences of internal oscillations. They emphasised the IPCC's exclusion of the millennial paleo-climatic data, and that the natural contributions associated with solar activity and internal variability could, in fact, be predominant during the recent warming period.

Many studies have concluded that the presence of oscillations that may affect Earth's climate is closely related to changes in solar activity over a wide range of timescales. Relationships between solar activity and the Earth's climate have been comprehensively reviewed (Haigh 1996, 2007; Gray et al. 2010). Kern et al. (2012) studied European lake sediments and reported the influences of 80-, 120-, 208-, 500-, 1000- and 1500-year cycles. They related these to solar activity. A multi-proxy study of the western Mediterranean region (Ramos-Román et al. 2018) extending over the entire 12,000 years of the Holocene showed the presence of millennial cycles attributed to external solar forcing, with a periodicity of 1430 years. Zhao and Feng (2015) found periodicities of 208, 521, and ~1000 years in Antarctic ice-core records, which they related to solar activity, particularly sunspot numbers. Raspopov et al. (2008) reported a clear ~200-year cycle for climate variations from Central Asia and related it to the solar de Vries/Suess cycle (de Vries 1958; Suess 1980).

The present study examines a limited number of published proxy temperature records for the Northern Hemisphere, including examples of both 'hockey stick' and MWP–LIA cycle profiles. The proxy records are first decomposed into sets of sine waves, which, when recombined, give the best simulation of the original temperature profile. This enables the periodicities of the main contributing sine waves to be compared.

The intention was to elucidate whether an examination of oscillatory behaviour can further explain differences between the hockey stick and the MWP–LIA cyclical representations. Decomposition into sets of oscillations can also be used to forecast temperatures into the industrial era, based on oscillatory patterns up to 1880 AD. The accuracy of fitting the data and simulating the natural patterns present can generally be improved by using the sine wave components as input to train a neural network (Abbot & Marohasy 2017a). This can potentially enable the attribution of temperature increases during the industrial era to either natural or to anthropogenic influences.

Proxy temperature records

There are hundreds of proxy temperature records reported in the literature, the majority corresponding to the Holocene period. For this investigation, five multi-proxy records for the Northern Hemisphere were selected for further analysis, each extending back over a period of at least 1000 years. These are a subset of the seventeen Northern Hemisphere records considered by Christiansen and Ljungqvist (2017) and include examples of both hockey stick and MWP–LIA cycle profiles.

The data for these five published Northern Hemisphere temperature proxy reconstructions were obtained from the National Oceanic and Atmospheric Administration, National Centres for Environmental Information Paleoclimatology Database (NOAA) (National Research Council 2006). Table 1.1 gives a summary of the proxy temperature reconstructions used for this analysis. The digital time-series were then examined by spectral analysis, using AutoSignal software, and applying the Parametric Interpretation and Prediction tools with Fourier Transform analysis.

In each case, the number of sine curves applied to reconstruct the total signal was increased until the fitting, estimated by a correlation coefficient, showed only marginal improvement. By adjusting the periodicities, and the phase and power of the identified sine wave components, the software optimises simulations of the original proxy signal, using a defined number of component sine waves. In each case, the optimisations were undertaken from the proxy record start date through to 1880 AD. This can be regarded as a pre-industrial period,

Table 1.1 Descriptions for the Northern Hemisphere multi-proxy temperature reconstructions used in this study

Proxy series	Date range	Proxy types	Number of records
Ljungqvist (2010)	800–1999	Extra-tropical historical documentary records, seafloor sediment records, lake sediment records, speleothem records, ice-core records, varved thickness sediment records, tree-ring width and maximum latewood density records	30
Moberg et al. (2005)	1–1979	Tree rings, lake and ocean sediments	11
Crowley and Lowery (2000)	1000–1993	Tree rings, ice cores, pollen, historical documents	12
Mann et al. (1999)	1000–1980	Tree rings	28
Schneider et al. (2015)	600–2002	Mid-latitude summer temperatures based on a wood density network	15

with only natural influences on climate, that is, without any anthropogenic (human-caused) contributions. The set of component sine waves obtained by spectral analysis can then be reconstructed into a composite and compared with the corresponding original temperature profile using the sine waves as input to a neural network.

Decomposition of proxy temperature records into sets of sine waves

Table 1.2 shows the results of spectral analysis for the five proxy temperature records for the Northern Hemisphere. Each of the individual temperature proxy records can be decomposed into a set comprising between four and ten sine waves, with periodic oscillations in the millennial, centennial and decadal ranges. For each sine wave, the periodicity is given in years and the power as a percentage. The power is proportional to the square of the amplitude of the sine wave.

Table 1.2 shows that the low-frequency components (millennial and centennial) make the major contribution to each composite proxy record, with decadal components making relatively minor contributions. In each case, there is a millennial frequency component falling in the range 978–1306 years.

Table 1.2 Component sine wave periodicities and power (%) for Northern Hemisphere proxy temperature records obtained using spectral analysis of data up to 1880 AD

	Millennial and centennial oscillations	Decadal oscillations
Ljungqvist (2010)	1230 (72%), 383 (11%), 149 (3%), 128 (4%), 106 (4%)	81 (2%), 76 (4%)
Moberg et al. (2005)	1223 (70%), 380 (10%), 183 (2%), 126 (3%), 106 (4%)	82 (2%), 75 (4%). 63 (2%), 55 (2%), 49 (2%)
Crowley and Lowery (2000)	1027 (72%), 538 (5%), 194 (6%), 171 (4%)	97 (2%), 83 (3%), 69 (5%)
Mann et al. (1998)	1306 (44%), 211 (10%), 119 (16%)	80 (4%), 66 (13%), 59 (4%), 40 (4%), 29 (5%)
Schneider et al. (2015)	1172 (19%), 537 (34%), 271 (4%), 187 (13%), 175 (14%)	56 (4%), 53 (5%), 49 (7%)

These results are in good agreement with the study by Lüdecke and Weiss (2017) who undertook a harmonic analysis of worldwide temperature proxies for 2000 years incorporating six previous global proxy temperature records. This showed the strongest components as sine waves, with periodicities of ~1000 years, ~460 years, and ~190 years, whereas other oscillations of the individual proxies are considerably weaker.

Humlum et al. (2011) undertook wavelet analysis showing the dominant periodic variations of 1130, 790–770, 560 and 390–360 years for Greenland ice-core records extending over a period of 4000 years. Zhao and Feng (2015) found periodicities of ~1000, 521 and 208 years in Antarctic ice-core records. Kern et al. (2012) studied European lake sediments and found the influences of millennial (1500 and 1000 years), centennial (500, 208 and 120 years) and decadal (80 years) cycles.

It is instructive to examine the Northern Hemisphere proxy reconstructions with regards to the dominant low-frequency oscillation (Table 1.2), and the subsequent development of either a hockey stick profile or MWP–LIA cycle. Figure 1.3 shows the dominant millennial sine wave component from spectral analysis of a hockey stick profile from Crowley and Lowery (2000) (red) and the MWP–LIA cycle from Ljungqvist (2010) (blue). It is evident that these two sine waves have similar periodicities (1027 and 1230 years respectively), equivalent amplitude or power (72%), and a closely corresponding phase.

Figure 1.4A shows the millennial sine wave from Crowley and Lowery (2000), as well as the individual centennial and decadal sine waves from spectral analysis. Figure 1.4B shows the millennial sine wave and the composite of the centennial and decadal sine waves, and Figure 1.4C shows the result of combining these.

Figure 1.5A shows the millennial sine wave from Ljungqvist (2010), as well as the individual centennial and decadal sine waves from spectral analysis. Figure 1.5B shows the millennial sine wave and the composite of the centennial and decadal sine waves, and Figure 1.5C shows the result of combining these.

Figure 1.3 Lowest frequency (millennial) sine wave component from decomposition of proxy temperature records

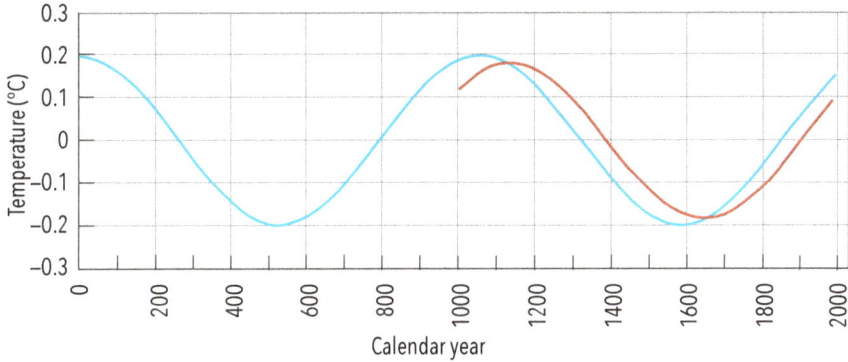

Source: Crowley and Lowery (2000) (red), and Ljungqvist (2010) (blue).

It can be concluded that starting from a similar millennial sine wave, through combination with an appropriate set of centennial and decadal sine waves, it is possible to generate either a hockey stick profile as illustrated by Crowley and Lowery (2000), or a MWP–LIA cyclic profile, as illustrated by Ljungqvist (2010).

The key difference in determining whether the profile develops into a hockey stick profile or MWP–LIA cycle appears to be the phase alignments, particularly of centennial cycles, with respect to the dominant millennial cycle.

Reconstruction of composite temperature signals

As demonstrated previously (Abbot & Marohasy 2017a) superior fitting to the original temperature proxies are generally obtained by using the individual sine wave components, and the composite as input data for a neural network, rather than applying a simple component addition. This data from the start of the proxy record up to 1880 AD serves as the training data for the neural network, enabling more complex mathematical use of the input data than a simple addition of the component sine wave oscillations. Application of the neural network to the input data generates better fitting to the training data set and this will generally

Figure 1.4 Sine wave components for Crowley and Lowery (2000)

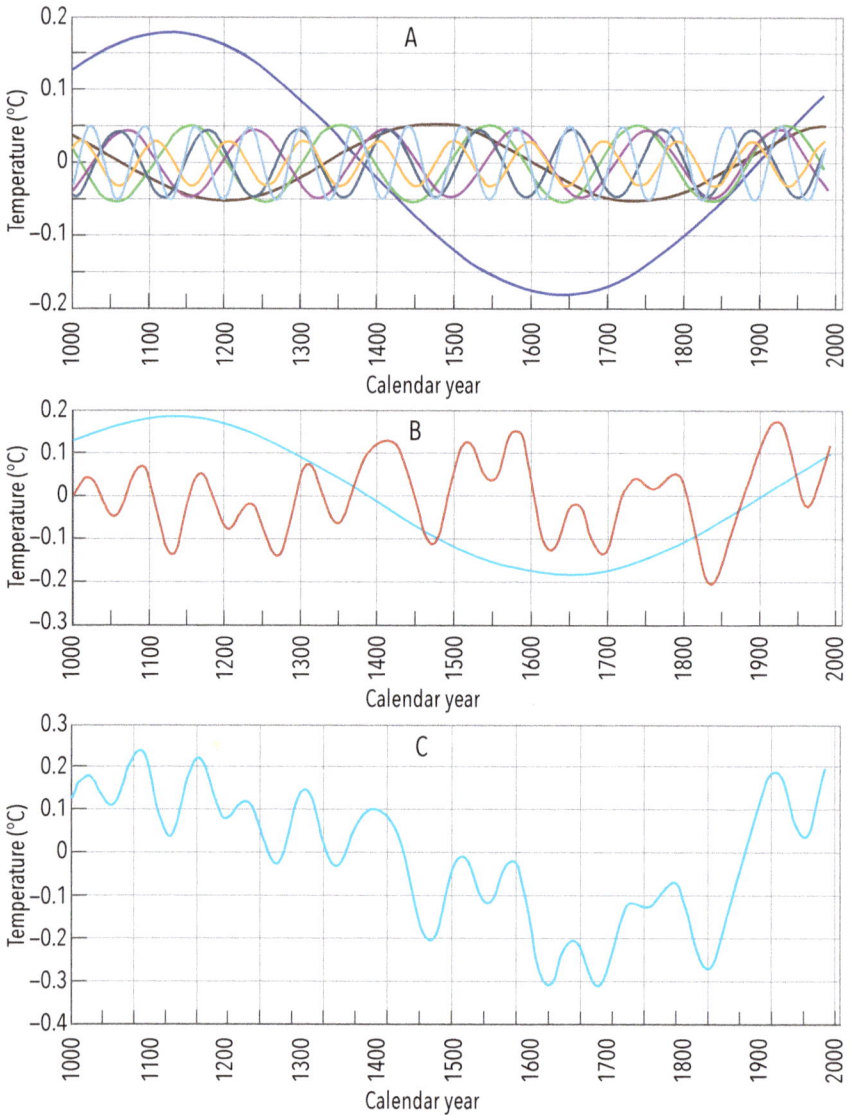

A: Millennial sine wave (dark blue) with individual centennial and decadal sine waves; B: Millennial sine wave (blue) with composite of centennial and decadal sine waves (red); C: Composite of millennial, centennial and decadal sine waves.

Source: Crowley and Lowery (2000).

Figure 1.5 Sine wave components for Ljungqvist (2010)

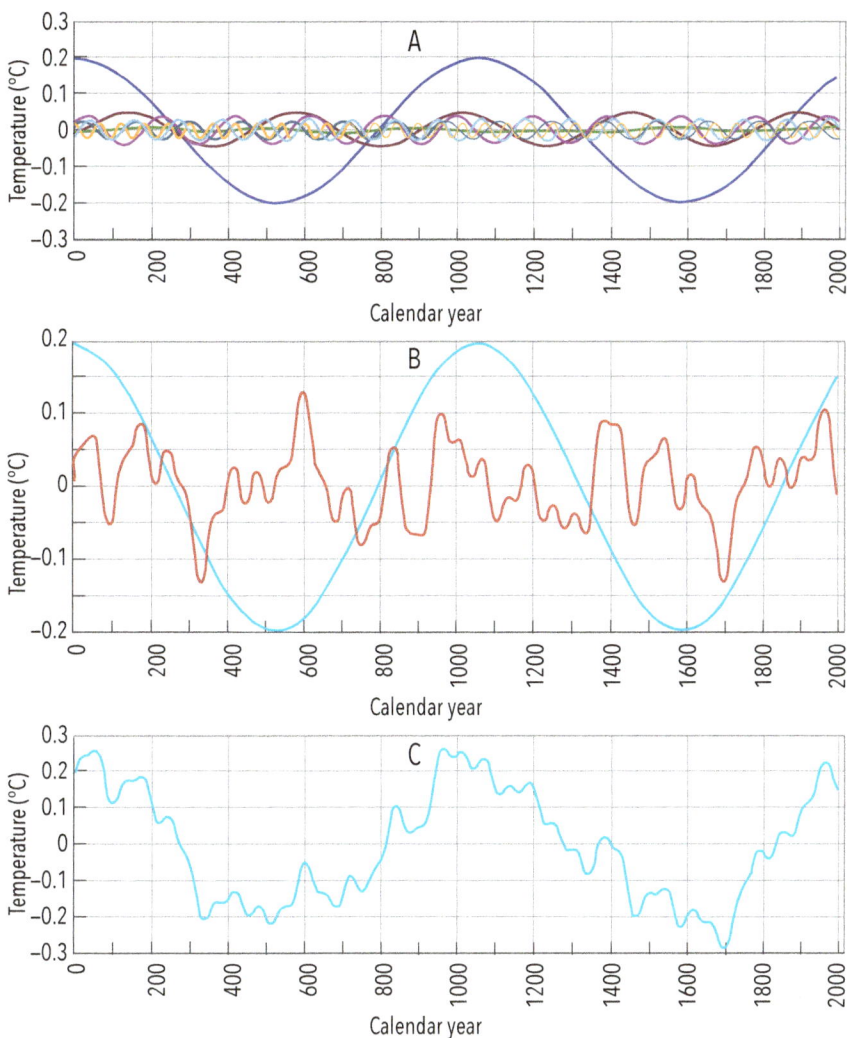

A: Millennial sine wave (dark blue) with individual centennial and decadal sine waves; B: Millennial sine wave (blue) with composite of centennial and decadal sine waves (red); C: Composite of millennial, centennial and decadal sine waves.

Source: Ljungqvist (2010).

lead to improved forecasting, assuming the relationships learned from the temperature data prior to 1880 AD are maintained into the industrial era.

Figures 1.6A through to 1.6E show the results of using sine waves from spectral analysis up to 1880 AD as input for training a neural network, enabling the projection of the results forwards into the twentieth century to the end of the respective proxy records. In each example, the upward warming trend beyond 1880 AD is present in the temperature forecasts. The temperature profiles based on oscillatory patterns learned by the neural network in the training process closely approximate to the actual proxy temperature records. Most of the warming can be explained in terms of a continuation of the natural oscillatory patterns present in the preceding 1000–2000 years.

It is also apparent that the predicted temperatures reached in the industrial era for the hockey stick profiles in Figure 1.6C (Crowley & Lowery 2000) and in Figure 1.6D (Mann et al. 1999) are higher than those in the proxy records of the previous 1000 years.

Figure 1.6A Proxy temperature profile reconstruction using sine wave output from spectral analysis as input to neural network

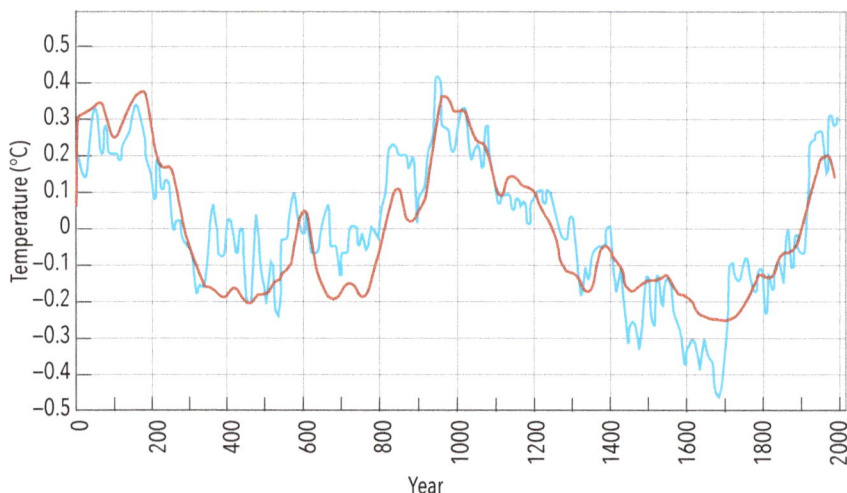

Proxy temperature profile (blue line) showing results from Ljungqvist (2010). Red line shows a reconstruction using an artificial neural network (ANN) with sine wave set from spectral analysis used as input. Data prior to 1880 AD used for ANN training and validation, and after 1880 AD for forecasting.
Source: Ljungqvist (2010).

Figure 1.6B Proxy temperature profile reconstruction using sine wave output from spectral analysis as input to neural network

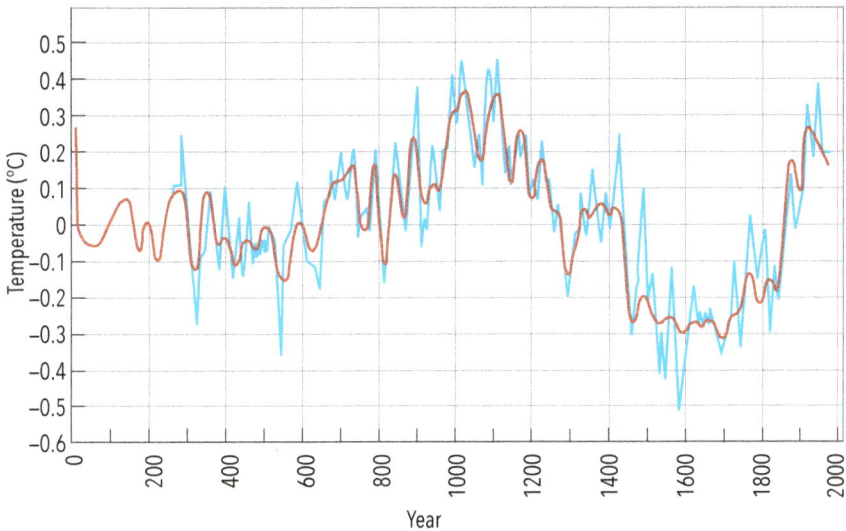

Proxy temperature profile (blue line) showing results from Mobherg et al. (2005). Red line shows a reconstruction using an artificial neural network (ANN) with sine wave set from spectral analysis used as input. Data prior to 1880 AD used for ANN training and validation, and after 1880 AD for forecasting (Abbot 2021).

Source: Moberg et al. (2005).

Figure 1.6C Proxy temperature profile reconstruction using sine wave output from spectral analysis as input to neural network

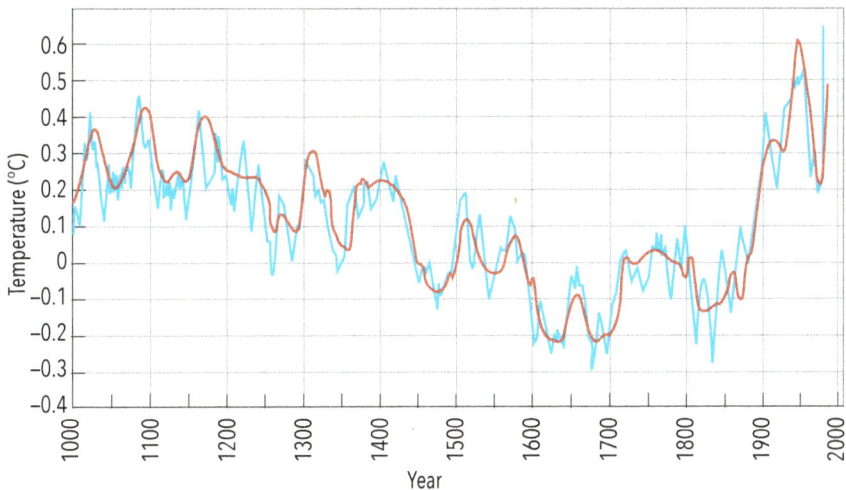

Proxy temperature profile (blue line) showing results from Crowley & Lowery (2000). Red line shows a reconstruction using an artificial neural network (ANN) with sine wave set from spectral analysis used as input. Data prior to 1880 AD used for ANN training and validation, and after 1880 AD for forecasting (Abbot 2021).

Source: from Crowley and Lowery (2000).

Figure 1.6D Proxy temperature profile reconstruction using sine wave output from spectral analysis as input to neural network

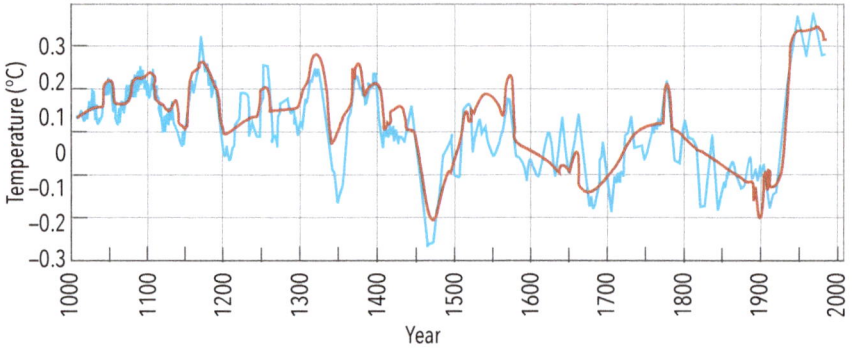

Proxy temperature profile (blue line) showing results from Mann et al. (1999). Red line shows a reconstruction using an artificial neural network (ANN) with sine wave set from spectral analysis used as input. Data prior to 1880 AD used for ANN training and validation, and after 1880 AD for forecasting (Abbot 2021).

Source: Mann et al. (1999).

Figure 1.6E Proxy temperature profile reconstruction using sine wave output from spectral analysis as input to neural network

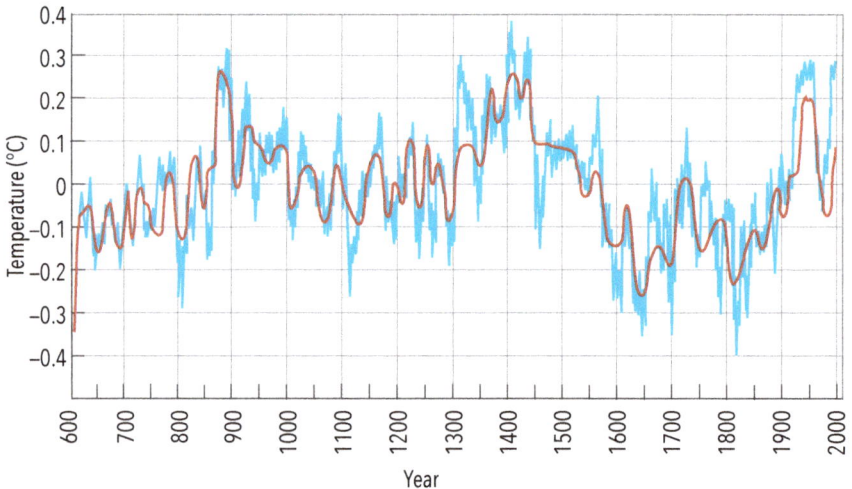

Proxy temperature profile (blue line) showing results from Schneider et al. (2015). Red line shows a reconstruction using an artificial neural network (ANN) with sine wave set from spectral analysis used as input. Data prior to 1880 AD used for ANN training and validation, and after 1880 AD for forecasting (Abbot 2021).

Source: Schneider et al. (2015).

Forecasting temperatures for the industrial era using identified oscillatory patterns

There have been comparatively few reported studies in which oscillatory patterns identified in proxy temperature records have been extended to make forecasts. One approach to temperature forecasting using oscillations is to use identified natural cycles such as the Pacific Decadal Oscillation (PDO) to make forecasts, as illustrated by Easterbrook (2016). An alternative approach is to mathematically examine the profiles of oscillatory behaviour in a temporal record, such as a typical proxy temperature record, in order to identify patterns that may continue into the future and, therefore, allow the prediction of future temperatures.

Humlum et al. (2011) examined proxy records for surface temperatures of Greenland extending back 4,000 years. Most of the dominant variations identified by the wavelet analysis are persistent with respect to strength and magnitude. Humlum et al. (2011) used three oscillations (2804, 1186 and 556 years) out of the ten identified by wavelet and Fourier analysis of surface temperatures of the Greenland ice sheet in order to project temperatures forwards from 1855 AD. The results were found to closely replicate the warm period over the past 150 years. The simple cyclic model was able to forecast the main features of this recorded warming until 2010, showing that a significant part of the twentieth century warming may be interpreted as the result of natural climatic variations, characterising at least the previous 4000 years.

In the present study, using Northern Hemisphere proxy temperatures, the spectral analysis was applied in each case from the start of the temperature record to 1880 AD. The extended profiles shown in Figures 1.6A to 1.6E from 1880 AD onwards are generated by extending the composite sine wave into the industrial era and are, therefore, forecasts. As demonstrated previously (Abbot & Marohasy 2017a), it is possible to improve the approximation of the proxy temperature profiles prior to 1880 AD by applying machine learning techniques, in particular artificial neural networks.

Oscillatory cycles in temperature records

Oscillatory processes are an important feature of the Earth's climate on a range of time frames including annual, decadal, centennial, and millennial (Bathiany et al. 2018). The presence of oscillatory processes is evident from visual inspection of most proxy and instrumental temperature records. However, it may not be apparent whether these are random oscillations, or the combined effects of underlying sets of oscillations with persistent periodicities, amplitudes and phase relationships. The presence of persistent, defined oscillations has been found in both instrumental and proxy temperature records using spectral analysis (Ludeke & Weiss 2017). For example, Wilson et al. (2007) examined tree-ring records covering a period of 1300 years for the Gulf of Alaska; several multi-decadal oscillations were identified by applying spectral analysis. Humlum et al. (2011) examined proxy records for surface temperatures of Greenland, extending back 4000 years. Wavelet analysis showed the dominant periodic oscillations at millennial (1130 years) and centennial (790–770 years, 560 years and 390–360 years) time frames. Most of the dominant oscillations identified by the wavelet analysis were found to be persistent with respect to strength and magnitude.

Millennial-scale periodicities

Abbot (2021) shows fourteen examples of millennial-scale periodicities reported from studies of proxy temperature records from different geographical regions and methods, including ice cores, tree rings, speleothems and lake sediments, with periodicities in the range 1067 to 1681 years. These can be compared with the range of millennial periodicities (978–1306 years) identified in the present study.

Millennial-scale climatic oscillations of ~1470 years are evident in numerous palaeo-climatic records in the North Atlantic and Pacific (Bond et al. 1997; Braun et al. 2005; Schulz 2002; Turney et al. 2004; De Menocal et al. 2000). Kelsey et al. (2015) discussed the existence of a ~1470-year cycle in the context of climate change. Bond et al. (1997; 1999) concluded from studies of the ratio of iron-stained to clean grains in ice-rafted debris in North Atlantic sediments, that climatic conditions have oscillated with an average period of about 1500 years over the past 100,000 years of the Holocene.

Centennial-scale periodicities

The present study shows the occurrence of centennial and multi-centennial oscillations for all eight proxy records examined in the range 106 to 538 years (see Table 1.2). Many other studies have also reported centennial-scale periodicities in temperature records. For example, Zhao and Feng (2015) found periodicities of 208 and 521 years in Antarctic ice-core records. Kern et al. (2012) studied European lake sediments and found the influences of 120-, 208- and 500-year cycles. Raspopov et al. (2008) reported the influence of ~200-year cycle on climate variations using results from the Central Asian Mountains. Liu et al. (2011) reported cycles of 800, 199, and 110 years in the tree rings of the Tibetan Plateau for the past 2,485 years.

Lüdecke et al. (2015a) used extensive northern hemispheric proxy data sets from Büntgen et al. (2011) and Christiansen and Ljungqvist (2012), together with a southern hemispheric tree-ring set, all with a one-year time resolution. Application of Fourier and wavelet transformation analysis, as well as nonlinear optimisation to sine functions, showed the dominance of a ~200-year cycle. Lüdecke et al. (2015b) found evidence for centennial cycles in proxy temperature records worldwide, ranging from Antarctica to Central Europe. Lüdecke and Weiss (2017) undertook a harmonic analysis of worldwide temperature proxies for 2000 years, incorporating six previous global proxy temperature records and reported centennial periodicities at ~460 years, and ~190 years.

For the Southern Hemisphere, (Cook et al. 2000) undertook spectral analysis of a Tasmanian tree-ring record of 3592 years. The reconstruction was dominated by oscillatory patterns of variability with centennial periods of 588 and 210–260 years.

Decadal-scale periodicities

Many studies have shown the significance of decadal and multi-decadal oscillations on temperature. For example, Semenov et al. (2010) described the impact of North Atlantic–Arctic multi-decadal variability on Northern Hemisphere surface air temperatures. The results stress the potential importance of natural internal multi-decadal variability originating in the North Atlantic–Arctic sector in generating inter-decadal climate changes, not only on a regional scale, but also possibly on a hemispheric and even a global scale. Dai et al. (2015) showed the need to incorporate natural oscillations, such as the IPO, to understand temperature variations in recent decades. Delworth et al. (2016) examined proxy reconstructions indicating substantial multi-decadal variability of the NAO over the past 1000 years and longer. They concluded that such NAO variations are likely to have altered the Atlantic Meridional Overturning Circulation (AMOC), and thereby influenced hemispheric-scale climate, in addition to the direct effect of NAO variations on atmospheric circulation and climate. Ortega et al. (2015) showed variations in NAO reconstructions for the past millennium using 48 proxy records.

Hernández et al. (2020) recently reviewed the significance of decadal oscillations throughout the Holocene, including ENSO, PDV, AMV, the NAO, the SAM and the IOD. Miller et al. (2018) attributed temperature changes in Maine, USA, to variations in the NAO and AMO over the past 900 years.

For the Southern Hemisphere, Cook et al. (2000) studied a Tasmanian tree-ring record of 3592 years. Spectral analysis of the temperature construction indicated that it is dominant by multi-decadal oscillatory patterns of variability with periods of 68–80 and 31 years.

Identifying features in proxy temperature records and the extent of their geographic locations

A major point of contention that has arisen when considering if Figure 1.1 or Figure 1.2 is a better representation of Northern Hemisphere temperatures is whether there are clearly identifiable MWP and LIA features in the temperature records during the past 1000 years (Broecker 2001). Soon and Baliunas (2003) presented a review of proxy climatic and environmental changes of the past 1000 years, examining more than 120 examples published between 1975 and 2002. These included regional and worldwide proxy records for both the Northern and

Southern Hemispheres. A total of 107 studies indicated there was a discernible climatic anomaly during the MWP (800 AD–1300 AD) in the proxy record, whereas only six indicated this was absent; 123 studies indicated there was an objectively discernible climatic anomaly during the LIA interval (1300 AD–1900 AD) in the proxy record, whereas only one reported this was absent.

Since 2003 there have been many additional proxy records published. For example, Dergachev and Raspopov (2010a, b) examined borehole data for Europe for the past 1000 years and concluded this clearly showed MWP and LIA features, in contrast to findings of Mann et al. (1999). McKay and Kaufman (2014) examined 59 individual proxy records including tree rings, lake sediments, marine sediments, glacier ice and speleothems extending back 2000 years for Alaska, Russia, Canada, Scandinavia and Greenland. These showed well-defined MWP, LIA, DA, and RWP features. A study by Auger et al. (2019) reported reconstructions for Arctic Canada over the past 2000 years showing MWP, LIA and RWP features. Pei et al. (2019) reported proxy-based temperature reconstruction for China during the Holocene exhibiting LIA, MWP and RWP features. Deng et al. (2017) showed distinct MWP and LIA periods from examination of sea surface temperatures from corals in the South China Sea over the past 1000 years.

Margaritelli et al. (2018) investigated the climate evolution of the last 2700 years in the central-western Mediterranean Sea, which they reconstructed from marine sediment records by integrating planktonic foraminifera and geochemical signals. The results provide the characterisation of climatic phases including the RWP, DA, MWP and LIA. Demezhko and Golovanova (2007) found evidence for the MWP and LIA for the Urals from 800 AD.

Lüning et al. (2019) have undertaken extensive studies of the Mediterranean region with regards to the MWP, examining trends from 79 published Mediterranean land and marine sites. They concluded that different regions have undergone warming or cooling during the past 150 years and attributed this to oscillatory patterns associated with AMO and NAO. They found that similar trends seem to have been developed during the MWP between 1000–1200 AD, when conditions in the western Mediterranean were generally warm (and dry), while large parts of the central and eastern Mediterranean were cold.

Luterbacher, et al. (2016) examined European summer land-temperature anomalies since 800 AD, which showed evidence for the MWP and LIA. The results show that subcontinental regions may undergo multi-decadal (and longer) periods of sustained temperature deviations from the continental average, indicating that internal variability of the climate system is particularly prominent at subcontinental scales.

Lüning et al. (2017) examined the MWP in Africa and Arabia and found evidence for both warming and cooling. The study was based on temperature records from 44 published localities. The vast majority of available Afro-Arabian onshore sites suggest a warm MWP, with the exception of the southern Levant (Israel, Palestine, and Jordan) where the MWP appears to have been cold. MWP cooling has also been documented in many segments of the circum-Africa–Arabian upwelling systems. Young et al. (2015) reported complementary paleoclimate proxy data suggesting that the western North Atlantic region remained cool, whereas the eastern North Atlantic region was comparatively warmer during the MWP – a dipole pattern compatible with a persistent positive phase of the NAO.

Considering the proxy temperature records of the Northern Hemisphere in the present study (shown in Figure 1.6), it is only in the case of Crowley and Lowery (2000) (Figure 1.6C) and Mann et al. (1999) (Figure 1.6D) – representing hockey sticks – that temperatures during the CWP are the maximum that have been reached.

Solar origin of oscillations

Huang et al. (2012) concluded that both the MWP and LIA are closely associated with the occurrence of solar activity over the past 1000 years through the effects of galactic cosmic rays. Cosmic rays are charged particles originating in outer space that bombard the earth's atmosphere. Studies show they may have an influence on cloud cover through formation of aerosols – tiny particles suspended in the air that seed cloud droplets. Other studies have discussed the connection of climate with cosmic rays and effects on cloud cover on timescales ranging from decadal to millennial (Usoskin & Kovaltsov 2008).

Solar activity, and its impact on the Earth's climate, is often related to total solar irradiance (TSI), which is a measure of the total energy received from the Sun at the top of the atmosphere (Dergachev & Volobuev 2018; Georgieva et al. 2015). Haigh (2007) reported that the absolute radiometers carried by satellites since the late 1970s have produced indisputable evidence that TSI varies systematically over the eleven-year sunspot cycle, but it is difficult to explain how the apparent response to the Sun, seen in many climate records, is directly attributable to these rather small changes in radiation. Haigh reviewed some of the evidence for a solar influence on the lower atmosphere and discussed mechanisms whereby the Sun may produce more significant impacts than might be surmised from a simplistic consideration only of direct variations in the magnitude of TSI.

Lüdecke et al. (2020) found evidence for significant influence of decadal and multi-decadal oscillations particularly the AMO and the NAO changes on European temperatures during the past century. They concluded that these oscillations are related to solar activity and have a direct correlation with the solar eleven-year Schwabe cycle. Van Geel and Ziegler (2013) discussed relationships between solar activity and decadal oscillations such as PDO and NAO.

Zhao, et al. (2020) found that the cross-wavelet correlations between the millennium-cycle components of sunspot number and Earth's climate changes remains both strong and stable during the past 8640 years (6755 BC–1885 AD). The Earth's climate indices exhibited the 1000-year oscillation corresponding to solar activity, which they associated with the Eddy cycle (Eddy 1976).

Kelsey et al. (2015) discussed the existence of an approximately 1470-year cycle associated with climate change and the possible origins of this cycle, which incorporates orbital, solar and lunar forcing.

Attribution of warming to natural and anthropogenic influences

The extent of relative attribution of warming during the industrial era to natural or anthropogenic influences is a complex issue that remains unresolved (Vitale & Bilancia 2013; Joos & Spahni 2008). It has been recognised that the inclusion of oscillatory processes is important when considering attribution (North 2012; Swanson et al. 2009). Some studies report that conclusions regarding the robustness of the finding that anthropogenic factors dominate are not affected by inclusion of oscillatory processes in climate models. However, these studies are often characterised by the limited selection of the short-term oscillatory processes that are included, often only decadal. For example, Imbers et al. (2013) included the ENSO or the AMO, while Mokhova and Smirnov (2018) included only a single oscillation, the AMO.

Gervais (2016), recognising the large amplitude of the natural 60-year cyclic component, questioned the occurrence of dangerous anthropogenic warming, which was found to be at most 0.6 °C once the natural component has been removed. Egorova et al. (2018) found that half of the simulated global warming is caused by the increase of GHGs, while the increase of the weakly absorbed solar irradiance is responsible for approximately one third of the total warming and did not directly account for oscillatory processes. Florides et al. (2010) concluded that, contrary to what has become common place, there are mechanisms other than atmospheric carbon dioxide concentration, such as solar radiation and galactic cosmic rays, which may be largely responsible for the observed changes in temperature.

Soon et al. (2015) concluded that most of the temperature trends since at least 1881 can be explained in terms of solar variability, with atmospheric greenhouse gas concentrations providing, at most, a minor contribution. The IPCC Fifth Assessment Report (Masson-Delmotte et al. 2013) concluded that solar activity related to TSI explains only a minor part of the temperature changes during the industrial era. However, others consider that direct changes in TSI may not fully represent the impact of solar influences (Haigh 2007).

General circulation models and oscillations

GCMs are mathematical models extensively used to represent the physical processes occurring in the atmosphere and ocean. They have been used to generate a predicted response of global climate to increasing greenhouse gas emissions (IPCC 2013; Tett et al. 2007; Santer et al.

2013; Zorita et al. 2005;). GCMs have been extensively used when attempting to differentiate between natural and anthropogenic global warming, for example in Eastern China (Li et al. 2007). Li et al. (2017) used GCMs to show that, between 1946 and 2005, GHG forcing was the primary driver of the surface warming over land in arid and semi-arid regions of the globe, and that the contribution from natural influences was negligible. Barkhordarian et al. (2018) used GCMs to conclude that the observed warming over northern South America between 1983 and 2012 was anthropogenic in origin. Allen et al. (2006) used GCMs to conclude that the anthropogenic influence is responsible for 0.3–0.5 °C warming per century during the twentieth century.

Several recent GCM studies have attempted to address their perceived shortcomings by incorporating natural oscillations. However, these studies are usually limited to considering one, or maybe several, identified natural oscillations, such as the AMO, NAO and PDO, which operate on decadal or multi-decadal timescales. Studies by Chylek et al. (2016) used GCMs and oscillatory processes to evaluate the relative contributions of natural and anthropogenic influences on temperatures during the period 1970–2005. Natural influences represented by AMO contributed warming between 0.13 and 0.20 °C compared to the GHG contribution of 0.49–0.58 °C. Wan et al. (2019) found that up to 0.5 °C of the observed warming trend may be associated with low-frequency variability of the climate, such as that represented by the PDO and NAO. Overall, the influence of both anthropogenic and natural external forcing is clear in Canada-wide mean and extreme temperatures. Folland et al. (2018) examined instrumental temperatures during the past century and found that GCMs are limited by the fact that they have not incorporated oscillatory modes over the past 100 years. New models were developed based on GCMs incorporating known oscillatory patterns including AMO, IPO, and ENSO. They concluded that both GHGs and natural oscillations are important, but the effects of GHGs dominate.

The present analysis shows that in terms of natural oscillations, decadal oscillations probably play a minor role compared to millennial and centennial oscillations, so that GCMs probably continue to underestimate the contribution of natural oscillations, which are apparent when considering timescales of 1000–2000 years.

Ongoing debate and possible resolution

Following the publication of MBH98 and MBH99 there has been intense and ongoing debate, both within the scientific community and outside, regarding the validity of the hockey stick, and the implications that would follow from its general acceptance. Several investigations (Zorita et al. 2003; von Storch et al. 2004; Bürger et al. 2006; Zorita et al. 2007) concluded that the methods applied would produce biased reconstructions with underestimated natural variability particularly in regards to lower frequencies. McIntyre and McKitrick (2003) concluded that the 'hockey stick' shape derived in the MBH98 proxy construction was primarily an artefact of poor data handling, obsolete data, and incorrect calculation of principal components. These criticisms were refuted by Wahl and Ammann (2007) suggesting errors in the methods used by McIntyre and McKitrick.

Mann et al. (2008) produced updated reconstructions of Earth's surface temperature for the past two millennia using a more diverse dataset that was significantly larger than the original tree-ring study, with over 1200 proxy records. They used two complementary methods, both of which resulted in a similar 'hockey stick' graph, with recent increases in Northern Hemisphere surface temperatures anomalous relative to at least the past 1300 years. Mann is quoted as saying, 'Ten years ago, the availability of data became quite sparse by the time you got back to 1000 AD, and what we had then was weighted towards tree-ring data; but now you can go back 1300 years without using tree-ring data at all and still get a verifiable conclusion' (Black 2008).

McIntyre and McKitrick (2009) perceived a number of problems with the Mann et al. (2008) paper, including some data with the axes upside down. Mann et al. (2009) replied that McIntyre and McKitrick 'raise no valid issues regarding our paper' and the 'claim that "upside down" data were used is bizarre', as the methods 'are insensitive to the sign of predictors'. They also said that excluding the contentious datasets has little effect on the result.

Esper et al. (2010) noted that more than two dozen large-scale climate reconstructions had been published, showing a broad consensus that there had been exceptional twentieth century warming after earlier climatic phases, notably the MWP and LIA. There were still issues of large-scale natural variability to be resolved, especially for the lowest frequency variations, stressing the need for further research. There is now broad recognition of the importance of preserving all possible climatic frequencies in proxy data and reconstructions. This is a step forwards when compared to simply using data that have likely not preserved low-frequency signals.

Christiansen and Ljungqvist (2017) presented an extensive review of large-scale temperature reconstructions of the past two millennia, particularly the variations in the methodologies for generating the types of proxy records reported in the literature, including those used in the present study. The different types of proxy records (historical documentary records, tree rings, ice cores, speleothems, terrestrial sediments, marine sediments, pollen, boreholes) were reviewed with particular emphasis on the possible limitations in their ability to capture low-frequency temperature information. The inherent complexity of the task is discussed with regards to many different factors that can influence the form of the resultant multi-proxy record including: the influence of number and positions of proxies; reconstruction methods; temperature proxy records and their limitations; and the influence of noisy proxies.

The apparent dichotomy between the types of multi-proxy temperature record may be a result of differences in reconstruction methods and selection of proxy records that are included (Christiansen & Ljungqvist 2017). It may also be explained in terms of the results of spectral analysis in the present study. Both the MWP–LIA cycle and hockey stick profiles can be interpreted as being based on a dominant millennial oscillation in the range 978–1306 years. Superimposed on this millennial oscillation are sets of centennial and decadal oscillations. The resultant effect of the centennial and decadal oscillations is significantly characterised by the phase shifts with respect to the millennial oscillation. In some cases, this reinforces the shape of millennial oscillation so that the distinctive MWP peak and trough of the LIA remain. In other cases, the result tends to counteract the peak and trough of the millennial oscillation, thereby flattening the appearance of the downwards trajectory between 1000 AD and 1850 AD and then reinforcing the rate of rapid warming from 1850 AD onwards. Different regions can experience warming or cooling at a particular time, as found for the Mediterranean during the MWP (Lüning et al. 2017). In a review of proxy temperature reconstructions over the past two millennia, Christiansen and Ljungqvist (2017) concluded that correlations between local temperatures and the Northern Hemisphere mean temperature are strongly geographically dependent. In particular, they found that the eastern Pacific and the northern North Atlantic show weak correlations with the Northern Hemisphere mean, while the interior of the continents show strong correlations. This may result in differences in phase alignment between the dominant millennial oscillation and centennial/decadal oscillations in different geographical locations. Therefore, the selection of proxy records weighted towards particular regions could introduce a bias towards a hockey stick or MWP–LIA cycle, enabling both to co-exist.

Acknowledgements

This research was funded by the B. Macfie Family Foundation.

This chapter is adapted from an article published in the journal *Earth Sciences*, available through open access and cited as: John Abbot. Using Oscillatory Processes in Northern Hemisphere Proxy Temperature Records to Forecast Industrial-era Temperatures. *Earth Sciences*. Vol. 10, No. 3, 2021, pp. 95–117. doi: 10.11648/j.earth.20211003.14

2 Climate Cycles Determined from the Ice Edge in the Barents Sea

Professor Jan-Erik Solheim

For millennia, the movement of the planets among the stars has been watched with curiosity. A long time ago it was discovered that Jupiter had an orbit of twelve years and Saturn had an orbit of 30 years. They meet in the sky every 20th year; after three meetings (which is five Jupiter periods, or two Saturn periods) 60 years has passed, and they are back to the same place relative the background stars. This remarkable synchronisation was called 'the music of the spheres' by Pythagoras of Samos (c. 570 – 490 BC).

As science evolved – with Johannes Kepler's elliptical orbits (published 1609–1619), Isaac Newton's mechanics (published 1687) – and the discovery by Christian Huygens (published 1665) that synchronisation between coupled oscillators needs very little energy exchange if enough time is allowed – we can also expect there to be synchronisation between the Sun and the planets, which have been in stable orbits for billions of years. Since short periods can be obscured by stochastic noise, it is important to find a long time series where long periods can be detected. In the following, I will demonstrate how a 442-year long time series of the ice edge in the Barents Sea may show a connection between the movement of planets and the Earth's climate.

Al Gore: 'The Arctic summer ice will disappear in 2014.' Did it?

In 2007, Al Gore and the International Panel of Climate Change (IPCC) received the Nobel Peace Prize. In his acceptance speech, Mr Gore proclaimed that the summer Arctic ice cap would be completely gone by 2014.

A few years earlier, in their yearbook for 1999, the Norwegian Polar Institute (NPI) published a map (Figure 2.1) that showed the maximum winter ice in the north Atlantic in four selected years in the month of April. During two of those years, 1769 and 1995, the ice edge was in the far north; and during the other two years, 1866 and 1966, it was further south. To these four extreme positions we have also added the maximum position of the ice edge in 2024 (ArticInfo). This map effectively demonstrates the

Figure 2.1 Extreme values of ice edge in the Barents Sea

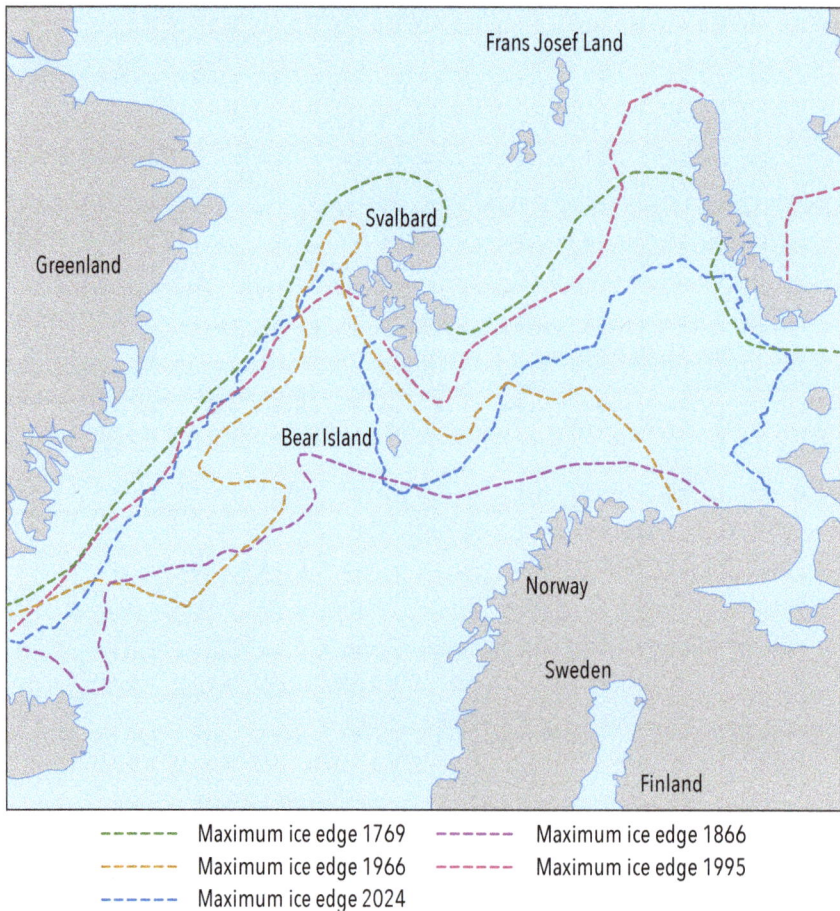

− − − − − Maximum ice edge 1769	− − − − − Maximum ice edge 1866
− − − − − Maximum ice edge 1966	− − − − − Maximum ice edge 1995
− − − − − Maximum ice edge 2024	

Source: NPI Yearbook 1999 and ArcticInfo 2024.

huge variations in ice coverage. However, the map that appeared in a pamphlet given to Al Gore – by the Norwegian Foreign Minister Jonas Gahr Støre (at the time of writing, October 2023, Støre is Norway's Prime Minister) and the Director of NPI Jan-Gunnar Winther – at the UN Climate Conference (COP 15) in Copenhagen in December 2009 was based on another publication that only covered the period from 1864 to 1998 (Vinje 2001). The curve that demonstrated the maximum winter ice edge in 1769 was erased. The pamphlet declared that the ice edge had retreated north from 1866 to 1995, following an increase in global carbon dioxide (CO_2) emissions.

Al Gore went on to predict that the summer ice would be gone by 2014 or 2016. But the ice maximum in 2024 showed more ice in the Eastern Barents Sea than in 1995.

What happened?
The National Snow and Ice Data Centre (NSIDC) publish daily, and monthly estimates of the ice cover based on satellite observations. In 2023, they published a September minimum of 4.23 million km^2. This is less than in 2022, but slightly more than in 2007. There is no trend in minimum Arctic ice in the period 2007 to 2021 (Astrup Jensen 2023). In the same period the atmospheric level of CO_2 increased from 385 to 420 ppm, or by about 9%.

We conclude that Al Gore and those who told him that the Artic summer ice would disappear in 2014 or 2016 were wrong. And the ice edge was further north between 1920 and 1940 than in 1966, so the graph given to Al Gore was factually wrong, even if it was politically correct.

To answer the question about the ice variations we need to study the more complete picture, which covers more of the last millennium: In order to present a better picture of the ice-cover development, a time series of 442 years of the summer ice edge is discussed in the following sections. But first we must consider the Barents Sea.

The Barents Sea and the Fram Strait

The west coast of Svalbard was discovered by Willem Barents in 1596. Barents noted that the sea teemed with whales and on the sea ice there were plentiful populations of seals. The whales were bowhead whales with large amounts of blubber (fat) from which oil could be extracted. Whaling expeditions (mostly from Holland) began off northwest Svalbard and the southern-eastern part of Fram Strait. The whaling industry grew into the first oil boom in Europe, creating 'summer cities' on Svalbard with many hundreds, maybe thousands, of people living there. The catch of the bowhead whale increased from about 1000 a year in around 1660 to a peak of 3000 in 1700. Thereafter, it fell to less than 500 a year up to 1820. The bowhead whale was declared nearly extinct a few years later.

Willem Barents had discovered an opening between Novaya Zemlya and a route into the Kara Sea on two earlier expeditions. In 1596, on his third expedition, he followed the ice edge east to the coast of Novaya Zemlya and then sailed north, hoping to circumnavigate the island. The ship became frozen in the ice, stranding Barents and his crew for the winter. Using two open boats, the crew finally reached the Murmansk area the following summer. Unfortunately, Willem Barents died before they made it back to Holland, but the ocean between Spitsbergen was named the Barents Sea in his honour.

The Barents Sea is an Arctic shelf sea, which in today's climate is partly free of ice, even during winter. The northward flowing Atlantic water that keeps the Barents Sea partly ice free, also keeps the Greenland Sea between Svalbard and Greenland mostly open during winter. Atlantic water is heavier (it has a higher density) than the fresh surface water from the ice melt, which keeps the freezing water temperatures in the upper 80 to 100 m of the sea's water column. In periods with little sea ice – such as between 1690 and 1780, and after about 1990 – warm, nutrient-rich water rises (upwells) to the surface during late winter and spring, increasing the biological production along the shelf break between Svalbard and Frans Josef Land (Falk-Petersen et al. 2000).

The ice in the Barents Sea consists of both newly frozen ice, and old sea ice transported with the trans-polar ice drift from the Kara Sea and Siberia. The first explorer to follow this ice drift was Fridtjof Nansen, sailing in the *Fram*. In 1893, the *Fram* became frozen in the ice at the

New Siberian Islands. Three years later the ship broke loose northwest of Svalbard – between Svalbard and Greenland, in what is now called the Fram Strait. Among many other lifetime achievements, Nansen also proved that the Arctic Ocean was more than 3000 m deep. The Nansen Basin in the Arctic Ocean is named after him.

In 1873, an Austrian–Hungarian expedition officially discovered Frans Josef Land, north of the Barents Sea, naming it after their emperor. It consists of 191 islands, mostly covered by permanent ice. The north-eastern part of Frans Josef Land is normally embedded in pack ice all year, but some years the pack ice retreats from the southern-most islands. An English trade route from London to Moscow, using ports along the northern coast of Russia was established from about 1555. This route

Figure 2.2 Barents Sea

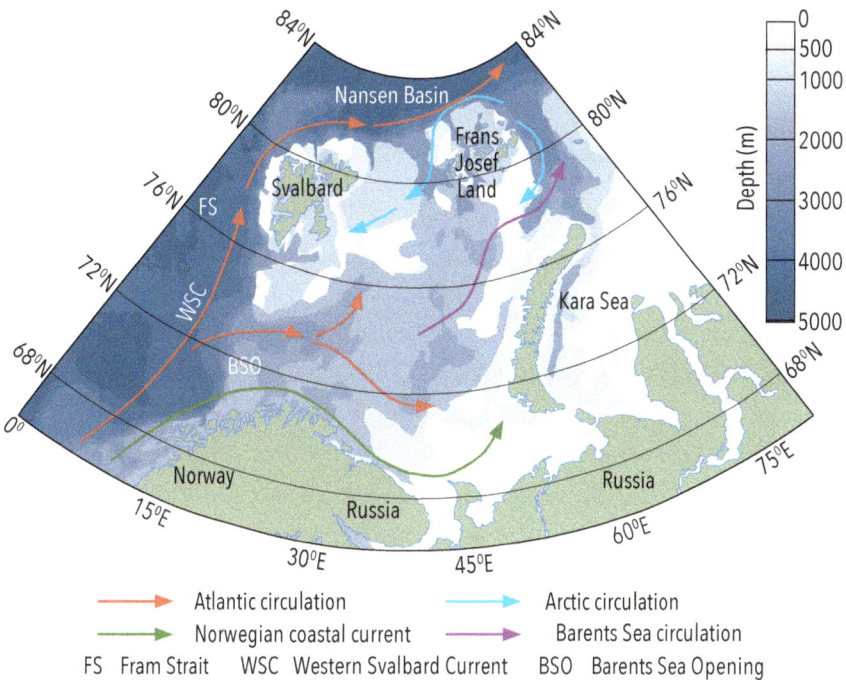

Atlantic circulation	Arctic circulation
Norwegian coastal current	Barents Sea circulation

FS   Fram Strait   WSC   Western Svalbard Current   BSO   Barents Sea Opening

The Barents Sea consists of a shallow shelf with much deeper water to the north: the Nansen Basin, and to the east: the Fram Strait. Circulation of main water masses is shown by arrows: red for the Atlantic water; blue for Arctic water; green for Norwegian coastal current and purple for Barents Sea water.

lasted 300 years and created a great deal of traffic in the Barents Sea and enormous wealth for the merchants in London.

A 400-year-long ice-edge series

In 1999, Torgny Vinje, one of the most important polar scientists in Norway during the 'modern polar era', published an estimate of the late August position of the ice edge in the Barents Sea during the last 400 years. This estimate was based on ice maps constructed from reports made by explorers, whalers, sealers, and fishermen, and from shipping activities in the Nordic Seas, including the Barents, Kara and White Seas. From 1950, he was able to use Russian, Norwegian, and American aircraft observations, which had become available, and from 1966 he could access satellite information. In addition to this data, information from the British merchants' journeys was used. This combined evidence resulted in a series that began in 1575, twenty years before Barents made his famous journey. Historic ice charts were collected and stored in the Arctic Climate System Studies (ACSYS) historial Ice Chart Archive (Løyning et al. 2003), which has now been taken over by NPI.

Vinje decided to estimate the average position of the ice edge in the sector between longitude 20 and 45°E, covering the eastern Barents Sea between Svalbard and Frans Josef Land during the two last weeks of August – the period when he found most data. An important question, though, is how to define the ice edge. The seal hunters operated as far into the ice as they could get. Sealers and whalers, who provided much of the data before 1950, reported ice concentrations from 30% to 60%, while wooden ships in the sailing area avoided concentrations greater than 30%; therefore, this became the conventional limit. However, in today's satellite era the ice edge has been redefined as 10% ice.

Vinje explained that, in addition to ordinary positional estimates, the ice-edge latitude is determined by sightings of a number of smaller islands in the sector, making the positions fairly reliable. However, he admits that he extrapolated from adjacent sectors or months in years when observations were not available in the selected sector, which was mainly before 1800. He claimed that this does not concern the most northern latitudes during the seventeenth century, which are direct August observations. He wrote:

> These old high latitude adventures have often been doubted, but the concurrence with a temperature optimum observed during the 17th century and the occurrence of repeated melt-back into the Arctic Ocean since 1935 contemporary with recent temperature increase, support the validity of these observations. It can be mentioned that the meridional circulation, or meridional exchange of heat, was relatively strong during the 17th century optimum period (Lamb 1995), and since 1970 we again observe a marked increase in the intensity of atmospheric circulation at higher latitudes.

Vinje (1999) found a strong correlation between the 10-year averages of the Northern Hemisphere mean temperature (NHMT) and Barents Sea ice-edge position (r^2 = 0.87) in August. He proposed that the ice-edge position could be used as a proxy for the NHMT, where 1° latitude in the position of the ice edge equated to approximately 0.2 °C in NHMT. From this he calculated a decrease in NHMT of 0.6 °C from 1580 to 1650, and a similar rise from 1650 to 1760, followed by a similar, rapid fall from 1760 to 1790, and finally an increase in NHMT of 0.7 °C from 1650 or 1790 until the 1990s.

A revised 442-year-long series

A team of us have expanded and revised the Barents ice-edge time series (Falk-Petersen et al. 2015; Mörner et al. 2020; Solheim et al. 2021). Satellite maps were used from 1979 and positions extrapolated from adjacent seas or months were removed.

This revised time series is presented as filled circles in Figure 2.3. The data set is not complete. From 1579 to 1678 the coverage is 39%. For the next hundred years, from 1679 to 1778, it declined to 34%, and between 1708 and 1722 there is a large gap. During the Little Ice Age (LIA), the late 1600s were the coldest in Europe. The coverage improved to 58% from 1779 to 1878, and to 66% in the following 100 years. From 1979, there is a complete coverage thanks to satellite observations. We also notice that the data vary less about a mean value from 1750.

Figure 2.3 The estimated position of the August ice edge in the Barents Sea since 1575

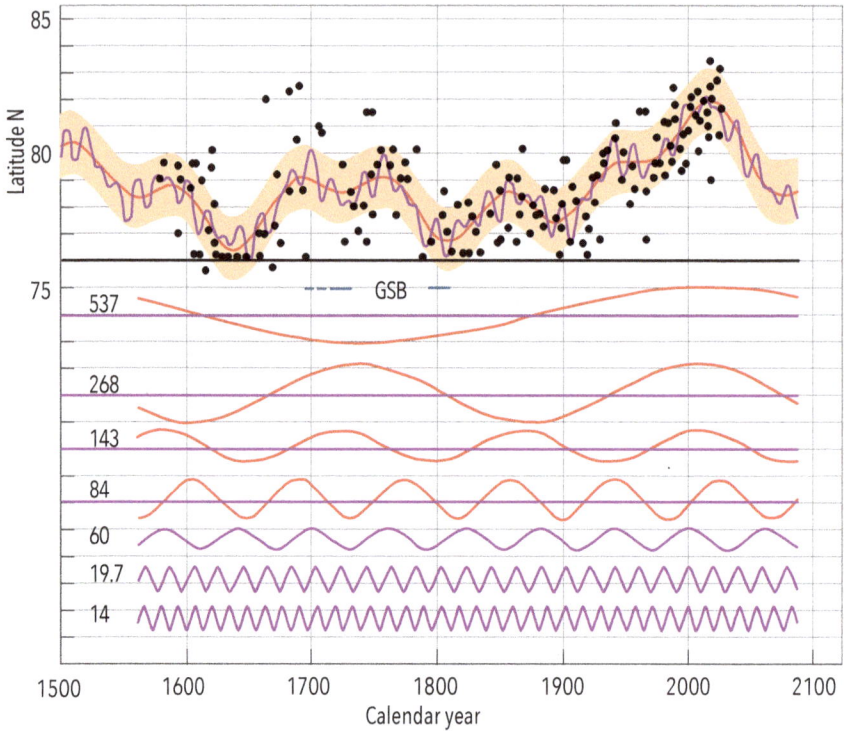

Positions of the August ice edge in the Barents Sea (filled circles) with a 4-periods model based on harmonics of planetary periods (red) with uncertainty interval (cream). The four periods (3, 3/2, 4/5)·P_{Jose} and P = 84 years are shown separately in the lower panels (purple). In addition, three low amplitude periods (P_{JS}, $3P_{JS}$ and $P_S/2$) are shown.

Source: Solheim et al. 2021.

And that the data had zero linear trend from 1579 until 1890. From 1890 to 2020 we found a trend of 0.035°N per year. This may represent the warming period after the LIA.

The estimated ice-edge positions vary between 75.5°N (1616, 1667–1668) and 83.35°N (2013). The data show two long minima: 1625–1662, and 1785–1812. The first corresponds to the early part of the solar Maunder Minimum 1640–1720 and the second to the beginning of the Dalton Minimum 1790–1820. However, several times between 1660 and 1710 we find ice-edge estimates north of 80°N. From

1850 the ice edge is almost always north of 78°N, except in the period from 1902 to 1917. After 1990, the August ice edge has always been north of 80°N, except in 2014. The data show no trend in the last ten years with an average value of 82.5°N. If we compare the four centuries 1600 to 2000, we find a binominal distribution with one peak at 76°N and one between 79 and 80°N in all centuries, except the 19th, which had a single peak at 77.5°N.

Fingerprints of the Sun and the planets

The giant planets of Jupiter and Saturn meet in the sky every 19.858 years. This is known as the Jupiter–Saturn synodic period (P_{JS}). This forces the Sun to move in a loop with approximately the same time period, disturbed by the position of the two other giant planets Uranus and Neptune. Jose (1965) was the first to notice, by undertaking detailed calculations, that the motion of the Sun about the centre of mass in the solar system (also called the barycentre) has a periodicity of 178.7 years, which is nine times the P_{JS}. He also found that solar cycles repeated with the same period – now called the Jose period (P_{Jose}). Charvátová and Hejda (2014) analysed the solar inertial orbit (SIM) and classified it as two types: an orderly type (trefoil in JS) that lasts about 50 years; and a disorderly type of variable length. During the orderly types of SIM we find solar maxima, and during disorderly phases we get deep minima. A summary of the SIM patterns is given by Charvátová and Hejda (2014). The Uranus–Neptune synodic period is 171.4 years. This makes the pattern change from one P_{Jose} to the next.

In Figure 2.3 we show how the August ice-edge position time series fits a series of harmonic periods. The four longest periods (red) have significant amplitudes. Three of them are harmonics of P_{Jose} and the fourth is the orbital period of Uranus (P_U). In addition, we determined three not statistically significant shorter periods shown as purple lines. They are P = 60 yrs = $3P_{JS}$; P = 19.7 yrs = P_{JS} and P = 14 yrs ≈ ½ P_S or $P_U/6$. Since all the periods are related to periods of the Jovian planets, and some are detected in solar variations (Yndestad & Solheim 2017), we have detected the fingerprints of the Sun and the planets. The four significant periods have maxima in around 2000, and it is these that are responsible for the increasing August ice latitude during the twentieth century. The longest period makes the peak around 2000 higher than the peak that was around 1750. The sum of the harmonic periods indicate that the August ice edge will move south during the next decade. This means there will be cooling in northwest Europe.

Synchronisation and Gulf Stream Beat (GSB)

Any curve can be modelled precisely with a number of harmonic functions (sinusoidals). However, such a model is only valid in the observed range. Only if the periods are related to stationary periods can the model be projected outside the observed range. This is the case here, where we have found planetary and solar periods where there are millions of years to synchronise.

An open question concerns the mechanism that forces the synchronisation. Our hypothesis is that the agents are gravitation from the planets and solar wind. The latter transfers angular momentum and the magnetic field, which interact with rotation of the Earth and ocean currents. For example, the Gulf Stream divides into a northern branch and a southern one. It is observed that when the Earth's rotation slows, the northern branch is stronger, and when it speeds up, the northern branch is weaker, causing a cold northern current to flow south along the coast of Western Europe. This is called Gulf Stream Beat (GSB), and this has happened several times during extended solar minima of the last millennium (Mörner et al. 2020) and it is indicated in Figure 2.3. We have presented a survey of possible driving forces in Mörner et al. (2020) and an analysis of the time series in Solheim et al. (2021).

Using Artificial Intelligence to Forecast Rainfall, Separation of Natural and Anthropogenic Components of Climate Change, and to Forecast Cyclone Trajectory and Intensity

3

Dr John Abbot

When floodwaters submerged much of Brisbane in January 2011, they drowned my much-loved red Corvette sports car in the basement car park of my unit in the riverside suburb of St Lucia. Compared to the heartache so many were experiencing, with the loss of their entire homes and tens of billions of dollars in damages to businesses, I had relatively little to complain about. The Corvette could not be salvaged. Its loss, and all the heartache that I could see across Brisbane, spurred me to develop a better technique for one-monthly rainfall forecasting using artificial intelligence (AI) with Jennifer Marohasy.

We spent the next few weeks compiling arrays of relevant climate data, including historic sea surface temperatures and pressures for the Pacific Ocean and rainfall for seventeen locations across Queensland. These arrays were fed into an artificial neural network (ANN) on my laptop, a program that I had been using for share market trading. We very soon realised that AI also had application for rainfall forecasting.

After we had a series of forecasts for all seventeen locations, we contacted the Australian Bureau of Meteorology (the Bureau). We asked for the equivalent forecasts from their simulation model, the Predictive Ocean Atmosphere Model for Australia (POAMA), in a form that would enable straightforward numerical comparisons to evaluate the skill (the measure of accuracy) of the two very different techniques for monthly rainfall forecasting.

This information was not forthcoming. The Bureau was only prepared to provide us with probabilistic forecasts, and not any direct output from their supercomputer. It was not until Marohasy, and I visited Melbourne in August 2011, and after protracted negotiations, that we got some data in a form that enabled us to publish our first paper comparing the one-monthly rainfall forecasts from the Bureau's general circulation model with output from our statistical model using AI.

We had hoped to develop a collaboration with the Bureau. The head of their long-range weather division, however, insisted AI had limited application for weather and climate forecasting because the climate was on a new and erratic trajectory.

In fact, we were a decade ahead of the times. It is now recognised that AI has much application for forecasting weather and climate. Marohasy and I went on to show how AI can also be used to separate out anthropogenic climate change from natural climate change, showing that natural influences are dominant in the industrial era.

Also included in this chapter is a summary of the reported application of AI to forecasting associated with cyclones, particularly relating to trajectory and intensity. A review of the literature from the past two decades shows that researchers from China and India have been particularly active in developing this approach.

Application of AI to weather forecasting

The application of AI techniques to forecasting weather and climate has resulted in hundreds of studies from many different parts of the world being published in the scientific literature over the past few decades. Many of these studies use AI techniques, particularly machine-learning (ML) methods such as ANNs. Data is used to train the ANN so that subsequent predictions can be made for a specified time ahead (lead time) and then compared with the actual value. By adjusting the architecture of the ANN and the data made available for training, an optimal forecasting program can potentially be developed. The accuracy of the forecast method for a specified variable, location, and lead time will depend on the availability of sufficient relevant input data and the existence of continuous relationships and patterns that the ANN can identify. This approach differs from the more traditional forecasting models that rely on the use of physical models. These assume that there is a good prior understanding of the underlying processes involved in the Earth's complex climatic system, and that these can be expressed mathematically. With the ML approach there is no requirement to impose an assumed set of relationships representing the Earth's extremely complex climatic system. The data speaks for itself; relationships that are presently unknown may be used to generate a skilful forecast. There is, of course no guarantee that patterns or relationships exist for any specified targeted forecast, but this will be revealed in the results obtained.

Despite the abundance of studies in the published literature showing good results for forecasting with AI, many national government agencies, including the Bureau, have been slow, or resistant, to adopt these new techniques, preferring to stay with their familiar physical models, even when these consistently produce poor results. The current media focus often confirms the more general fears that may exist in many traditional areas, where predominantly human activities may potentially be replaced with superior performing AI.

Current media attention on AI has included a focus on applications for weather forecasting. One prominent study receiving attention, published by Lam et al. in *Science* in November 2023 entitled, 'Learning skillful medium-range global weather forecasting', was extensively reported in the media, for example in the *Washington Post*, the *Financial Times* and on the BBC *News*. They report that traditional numerical weather prediction uses increased computer resources to improve forecast accuracy, but cannot directly use historical weather data to improve the underlying model. Lam et al. (2023) introduced a machine-learning–based method called 'GraphCast, which can be trained directly from data. This enabled prediction of hundreds of weather variables, globally, over ten days, in less than one minute. GraphCast significantly outperforms the most accurate operational deterministic systems on 90% of 1380 verification targets, and its forecasts give better severe event prediction, including tropical cyclones (TCs), atmospheric rivers, and extreme temperatures.

An example of a successful application of AI to weather forecasting is the prediction of monthly rainfall at long lead times, as pioneered with Marohasy, beginning in 2011 after the flooding of Brisbane.

Daily rainfall data has been collected by the Bureau at many sites in Australia over many decades, and some sites have data extending back more than 100 years. Using this data as input, combined with other data including local temperatures and climate indices, it has been possible to generate rainfall forecasts with lead times ranging from three months to several years (Abbot & Marohasy, 2012, 2013, 2014, 2015a, b, c, d, 2016a, b, 2017a, b, 2018). The skill of these forecasts can be quantitatively compared with forecasts generated using physical models, such as those used by the Bureau. The studies show that ANNs forecast with long lead times, between one and five years, significantly more skilfully than equivalent forecasts using physical models.

Considering the town of Miles in Queensland as an example, using an ANN it is possible to show a good correspondence between ANN forecasts at twelve months lead time and the actual monthly rainfall (see Figure 3.1). Table 3.1 shows a comparison of skill score for Miles using AI and forecasts using a physical model. The skill scores are calculated relative to

Figure 3.1 Forecast and observed monthly rainfall for Miles using the composite neural network. Test period August 2004 to July 2011

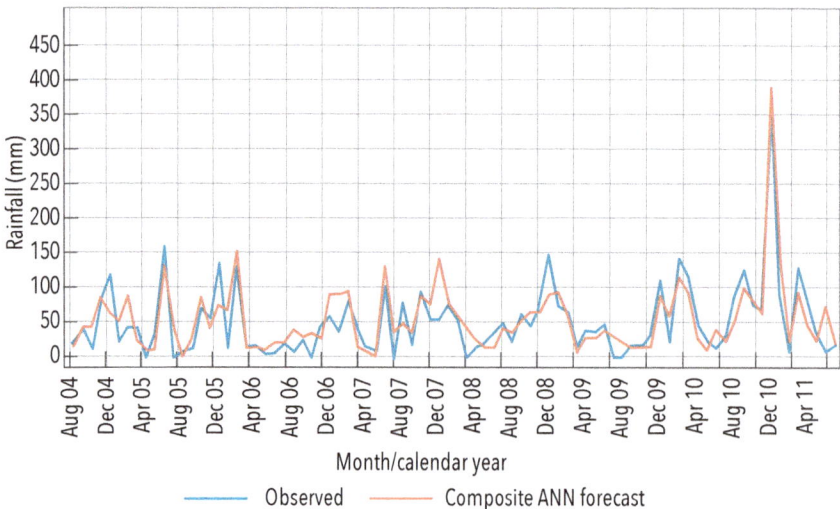

Table 3.1 Forecast skill scores for monthly rainfall forecasts for Miles with twelve months lead

Month	Skill score (%)	
	ANN (lead 12 months)	POAMA (lead 8 months)
January	30.2	6.0
February	22.2	−1.7
March	49.3	−6.3
April	39.3	−40.6
May	33.1	−5.9
June	43.3	−1.2
July	62.1	−3.4
August	30.6	0.0
September	69.0	−2.0
October	43.0	−8.0
November	22.0	8.1
December	48.6	−16.1
Average	41.1	−5.9

climatology, which is a simple long-term monthly average over 30 years. A skill score of 0% represents no improvement over climatology, while 100% is a perfect forecast. Table 3.1 shows the AI forecasts are more skilful than climatology in every case, while the POAMA physical models that run on very expensive supercomputers are on average less skilful than simply taking a long-term average for each month.

Oscillations in temperature records

Instrumental temperature records in many parts of the world generally extend back to a little over a century. To understand how climate has varied over much longer periods, such as over hundreds and thousands of years, various types of proxy records have been assembled. These are derived from measurements associated with biological and geological phenomena that can leave evidence of past climate, particularly of temperatures. There are hundreds of proxy temperature records reported in the scientific literature corresponding to the Holocene period – which is the last approximately 10,000 years. The most familiar proxy records are derived from annual rings of long-lived tree species. Other proxies include measurements from corals, stalagmites, and sediments. These types of records provide evidence for periods of time over the past several thousand years (the late Holocene) that were either colder or experienced similar temperatures to the present, for example, the Little Ice Age (LIA) and the Medieval Warm Period (MWP) (Hunt 2006; Lamb 1965; Lamb 1982).

For example, Lüdecke et al. (2013) examined six long European instrumental temperature records of monthly averages reaching back as far as 1757 AD. Reliable monthly temperature records exist for Prague (in the Czech Republic), Hohenpeißenberg and Munich (in Germany), Kremsmunster and Vienna (in Austria), and Paris (in France). All six temperature profiles exhibit an approximate V-shape, with temperatures decreasing during the nineteenth and rising in the twentieth century. The temperature records from the six Central European stations show remarkable agreement, justifying their averaging to produce a Central European temperature record. Spectral analysis generated six strong periodic components at 254, 85, 64, 51, 42 and 36 years. Very good agreement of temperature reconstruction and the original profile was found using these six components, showing that the present climate dynamics are dominated by periodic processes. Lüdecke and Weiss (2017) investigated the dominant cycles in worldwide temperature proxy data for the last 2000 years, complemented by instrumental temperature measurements of global temperatures, with the strongest components at ~1000-, ~460-, and ~190-year periods, most likely having a solar origin.

Using AI and component oscillations to forecast natural temperature changes

Examination of many proxy temperature records shows they typically consist of complex oscillations or cycles, with the amplitude and structure of the temperature signal depending on the geographical location considered. In the pre-industrial era, these oscillations represented the compound effect of natural phenomena, both internal (e.g. North

Atlantic Oscillation (NOA), El Niño Southern Oscillation (ENSO)) and external (for example, solar and volcanic activity). With the growth of industrialisation since about the mid-nineteenth century, there is the possibility that there is also a contribution to climate change from anthropogenic greenhouse gases, particularly carbon dioxide (CO_2) and methane (CH_4). However, the relative contributions of natural cycles and anthropogenic effects is far from certain (van Geel & Ziegler 2013) and there is continuing interest in attempting to answer this question of evaluating the relative contributions (Andres & Peltier 2015; Gervais 2016; McShane & Wyner 2011; Moberg 2013).

Examples of proxy records are shown below (blue line) for the Northern Hemisphere (Figure 3.2) using results from Esper et al. (2002), and for regional southern South America multi-proxy temperatures (Figure 3.3) (Neukom et al. 2011). The presence of a relatively warm period at around 1000 AD followed by colder period (1400–1600 AD) is evident. This agrees with the occurrence of the MWP and LIA clearly apparent in the IPCC First Assessment Report (1990). In later IPCC reports, however, this was replaced with the 'hockey stick' profile where these features are not clearly present and there is a slow gradual decline in temperature for about 1000 years, followed by an abrupt increase during the industrial era.

The red lines in Figures 3.2 and 3.3 show the results from decomposing the original data into a set of component sine wave oscillations, then generating a temperature profile using the component oscillations as input to a neural network (Abbot 2021; Abbot & Marohasy 2017c). The component oscillations for Figure 3.2 corresponded to periods of 1173, 460,192, 113, 69 and 49 years. The component oscillations for Figure 3.3 corresponded to period of 745, 383, 245, 217, 141, 129, 89, and 52 years. A composite signal can be constructed from the sets of component sine waves through the simple addition of the sinusoidal components. This composite signal can itself be used as the basis of making projections of temperature. However, superior fitting to the temperature proxies are obtained by using the sine wave components and composite as ANN input data (Abbot 2021; Abbot & Marohasy 2017c).

Figure 3.2 Proxy temperature profile (blue) and neural network output (orange) for the Northern Hemisphere

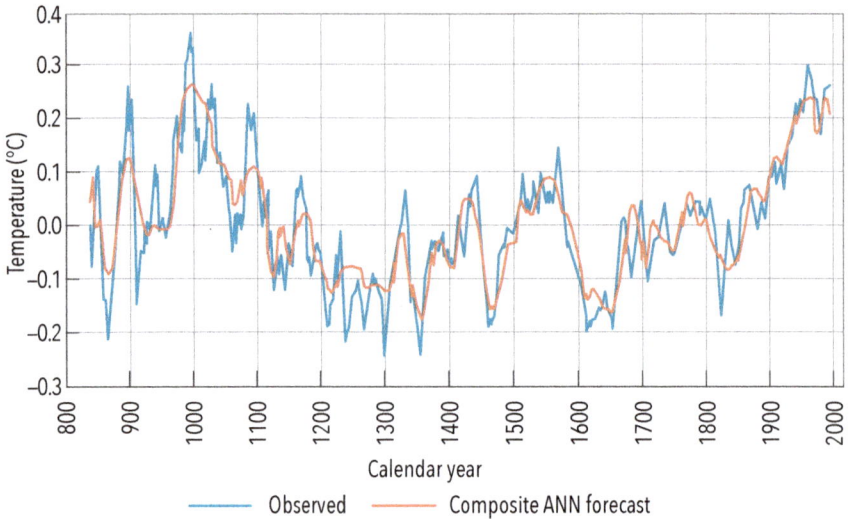

Source: Using results from Esper et al. (2002). ANN forecast shown from 1880 AD corresponding to industrial era.

Figure 3.3 Proxy temperature record (blue) and neural network output (orange) based on input from spectral analysis for regional southern South America multi-proxy temperatures

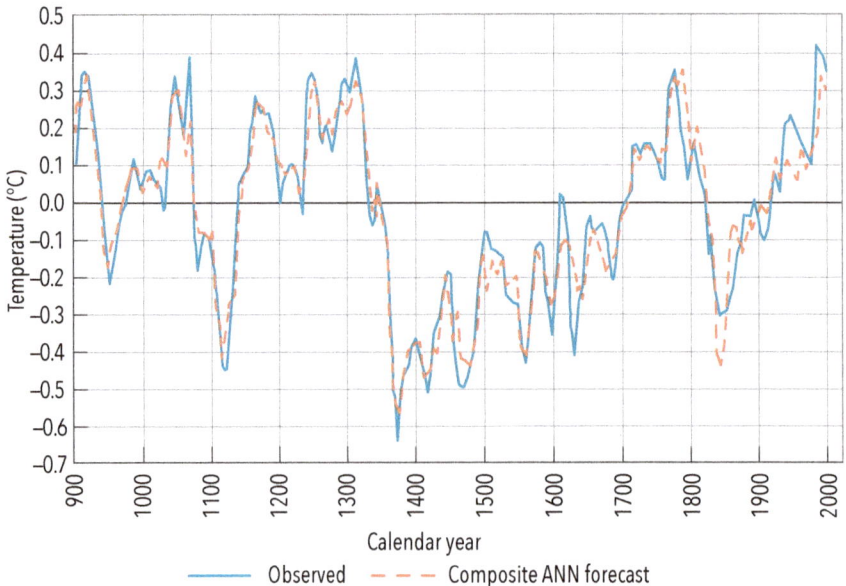

Source: Neukom et al. (2011). ANN forecast shown from 1880 AD corresponding to industrial era.

The ANN was trained and validated using data up to 1880 AD. It can be assumed that during this period of approximately 1000 years the anthropogenic contribution was minimal and it is regarded as representing natural phenomena affecting the climate. The red lines after 1880 AD were generated by using the trained ANNs and extended oscillations as input data to forecast the profiles from only natural contributions to temperature change. It can be seen that in both cases there is an upward trend in temperature between 1880 and 2000 AD, and small deviations between the red and blue lines. This implies that the increase in temperature after 1880 AD was predominantly due to a continuation of the influences of natural factors also present during the pre-industrial period.

Equilibrium Climate Sensitivity

The Equilibrium Climate Sensitivity (ECS) refers to the equilibrium change in global mean near-surface air temperature that would result from a sustained doubling of the atmospheric (equivalent) CO_2 concentration. It is derived from the general circulation models (GCMs), together with a set of external radiative forcing (RF) functions (Andrews et al. 2012; Flato et al. 2013; Myhre et al. 2013). The external RF functions were initially based on the work of Arrhenius in the nineteenth century, who first proposed that a doubling of atmospheric CO_2 could lead to a 5 to 6 °C increase in global temperatures (Arrhenius 1896). These estimates have since been progressively revised down, but are generally much larger than estimates calculated from experimental spectroscopy lying in the range 0.3–0.9 °C. Current estimates of the ECS using GCMs lie in the range 2–4.5 °C (Flato et al. 2013).

Based on the ANN methods described here, an ECS in the range 0.6–0.8 °C can be calculated. This is significantly lower than values reported using GCMs because it recognises the contribution from continuation of natural cycles. Furthermore, this should be regarded as the maximum value associated with anthropogenic emissions. There is evidence that rising temperatures lead CO_2 concentration increases. Natural temperature increases can be associated with warming oceans liberating CO_2 into the atmosphere.

Deficiencies with GCMs

A major limitation of GCMs is that they do not adequately generate the necessary amplitude of temperature oscillations apparent in proxy records extending back over several millennia (Scafetta 2013). The relatively small attribution to natural phenomena leads to temperature profiles typified by the "hockey stick" (Mann et al. 2008). However, the majority of proxy temperature profiles representing the past several millennia found in the scientific literature, whether local, regional, or global, do not resemble this hockey stick profile, but rather exhibit significant amplitude in oscillations over the past several millennia. For example, many regional temperature proxies for the Northern Hemisphere show pronounced cyclical behaviour corresponding to the MWP and the LIA (Guangliang et al. 2013; Ljungqvist et al. 2012; Priem 1997; Taricco et al. 2015) as illustrated in Figure 3.2.

Several recent GCM studies have attempted to address perceived shortcomings by incorporating natural oscillations. However, these studies are usually limited to consideration of one or several identified natural oscillations, such as the Atlantic Multi-decadal Oscillation (AMO), the NAO and Pacific Decadal Oscillation (PDO) that operate on decadal or multi-decadal time scales. Studies using GCMs and oscillatory processes to evaluate the relative contributions of natural and anthropogenic influences on temperatures during the period 1970 to 2005, (Chylek et al. 2016) showed natural influences represented by the AMO contributed warming between 0.13 and 0.20°C compared to the GHG contribution of 0.49–0.58 °C. Another study (Wan et al. 2019) found that up to 0.5 °C of the observed warming trend may be associated with low frequency variability of the climate, such as that represented by the PDO and NAO.

Decadal-scale oscillations probably play a minor role compared to millennial and centennial oscillations, so that GCMs probably continue to underestimate the contribution of natural oscillations that are apparent when considering time scales of 1000 to 2000 years.

AI forecasting of cyclone trajectory and landfall location

Tropical cyclones (TCs) are one of the most severe meteorological phenomena, therefore, making rapid and accurate track forecasts is crucial for disaster prevention and mitigation (Wang et al. 2023). Because TC tracks are affected by various factors (the steering flow, the thermal structure of the underlying surface, and the atmospheric circulation), their trajectories exhibit highly complex nonlinear behaviour.

Traditionally, track forecasting methods have primarily relied on Numerical Weather Prediction (NWP), which is extensively employed by official meteorological organisations (Geng et al. 2023). NWP models entail complex large-scale dynamics and physics calculations for forecasting, requiring significant computational resources, including supercomputers. To address this challenge, researchers are exploring alternative, more efficient and accurate methods such as approaches incorporating AI.

Roy and Kovordányi (2012) reviewed the application of ANNs in forecasting the tracks of TCs, which are governed by a range of factors including weather conditions, wind pressure, sea surface temperature, air temperature, ocean currents, and the Earth's rotational force. They noted that combining these parameters to produce reliable and accurate forecasts is a formidable task. However, in recent years, with the increasing availability of suitable data, more advanced forecasting techniques have been developed, particularly using ANNs.

Kovordányi and Roy (2009) have discussed methods for forecasting cyclone tracks using ANNs with satellite images as input data. In their study, a multi-layer ANN, resembling the human visual system, was trained to forecast the movement of cyclones based on satellite images. The trained network produced correct directional forecast for 98% of test images, thus showing a good generalisation capability. The results indicate that multi-layer ANNs could be further developed into an effective tool for cyclone track forecasting using various types of remote sensing data.

The north-western Pacific is the most active basin for TCs in the world, generating more than one-third of the total number of TCs. China, located on the western side of the Pacific Ocean, with a coastline longer than 18,000 km, is severely influenced by TCs. These storm systems are accompanied by strong winds, heavy precipitation, and storm surges, resulting in severe disasters that affect human lives and economic growth. ML techniques, particularly NNs, have been applied to produce track forecasts in this region of the world (Cheung et al. 2021; Tan et al. 2021).

AI forecasting of cyclone intensity, winds, storm surge

A number of studies published in past decade show that ML and particularly ANNs can significantly enhance the forecasting of cyclone intensity (Chen et al. 2019; Wu et al. 2022). The study by Ko et al. (2023) focused on developing a ML model for TC intensity change forecasting, especially for rapid intensification (RI). The ML model used input data extracted from a high-resolution hurricane model, which contained 21 or 34 RI-related predictors. Chen et al. (2023) used NNs for predicting TC RI. Compared with the operational forecasts provided for western Pacific TCs, the results achieved higher RI detection probabilities and lower false-alarm rates.

Na et al. (2022) generated near real-time predictions of TC intensity in the north-western Pacific Ocean using NNs. Factors predicted included maximum sustained wind speed, and minimum sea-level pressure. The accuracy of the six- to 24-hour forecasts by the model was comparable or better than other methods. Chen et al. (2022) used NNs to develop a model for TC RI forecasting and showed that these exhibited superior results compared to operational forecasting methods. The model showed a 53.3% improvement in the probability of detection, and a 42.9% reduction in false alarms. Lin et al, (2023) applied NN models for TCs in the western Pacific and produced forecasts better than many existing numerical forecasting methods, statistical forecasting methods, and ML methods.

Sarkar et al. (2022) used NNs for forecasting pressure drop and maximum sustained wind speed (MSWS) associated with cyclonic systems over the Bay of Bengal. The results illustrate forecasting the MSWS during mature stage of cyclonic structures, with the lowest prediction error at a 60-hour lead time. Results were compared with the operational forecasts for ten cyclonic systems from 2016 to 2019. Snaiki and Wu (2022) used NNs for simulation of extratropical cyclone wind risk applied to eastern North America. Kolukula and Murty (2022) investigated improving cyclone wind fields using NNs and their application in extreme events. Sahoo and Bhaskaran (2019) investigated the prediction of storm surge and inundation using climatological datasets for the Indian coast, particularly in the Bay of Bengal region, using ANNs.

Conclusion

The uptake of AI methods in climate science, specifically ANNs, has generally been slow by official agencies such as the Bureau in Australia. This is in part due to the heavy investment in physical models, particularly GCMs, and the importance of these to the theory of anthropogenic global warming. However, the complexity of the climate systems and the current, limited understanding of all the physical processes leads to large uncertainties in the results generated, including the ECS, which indicated the atmospheric temperature increase due to a doubling of CO_2 concentration. My work with Marohasy forecasting monthly rainfall for locations in Queensland, and also global temperatures for the twentieth century, show the application of AI for both weather and climate.

Our work agrees with other studies that stress the importance of acknowledging the presence of oscillatory processes that operate at a range of time scales, and which have persisted over thousands of years.

There is merit in attempting to construct physical models to help with the understanding of climatic phenomena. However, the inherent complexity of the Earth's climatic system should lead to caution in the confidence of model outputs, particularly when important characteristics such as oscillatory processes are not well replicated in outputs. An alternative to this approach, as demonstrated here, does not require a prior understanding of the physical processes, but does require adequate relevant data and the appropriate application of AI techniques. This approach may also eventually lead to better understanding of the physical processes through the identification of important cyclical phenomena and their interactions. AI has been shown to consistently outperform the physical models, whether it is for forecasting seasonal rainfall, or extreme weather events such as hurricanes. The adoption of AI for better management of water infrastructure should become mandatory, ensuring that Brisbane is never flooded again.

4 The Geology of Climate Change

Emeritus Professor Ian Plimer

Climate change is complex. It is only possible to attempt to understand climate if all the data from the spectrum of earth and planetary sciences is integrated. Planet Earth is in the Goldilocks zone where liquid water has existed for billions of years and, unlike other planets, does not have diurnal temperature extremes, thereby making Earth habitable. Notwithstanding, variation in temperature over the course of human habitation has been from icehouse to hothouse conditions, no different from modern latitudinal climate variation.

First order climate changes result from changes in astronomical forcing (Milankovitch cycles) and solar activity; second order climate changes are related to plate tectonics. These first and second order climate changes cannot be explained by changes in atmospheric carbon dioxide (CO_2). If third order climate changes are to be valid, then it must be shown that human emissions of CO_2 drive global warming. This has not been done. However, heat generation by cities and water cycle changes have been shown to be of human origin and can affect local weather and local climate.

It is worth making the point that for more than 80% of geological time there has been no ice on Earth, the planet was warmer and wetter, and sea levels higher. The geological history of the Earth shows that the major processes driving climate change are orbital cycles, solar activity, cosmic and galactic effects, and plate tectonic processes, such as mountain building and volcanism. There is no reason why these major planetary processes should be different because humans currently exist on Earth. The current climate is not unprecedented. It is driven by the same cyclical and random processes that have operated for billions of years.

This chapter examines these forces and their effects on planet Earth's climate.

Tectonic climate cycles

Long-term climate cycles are in the order of 400 million years, when supercontinents are pulled apart and the fragments stitched back together. At these times, the carbon and oxygen cycles operate in tandem and may well drive evolution and extinction (Large et al. 2019). Supercontinent fragmentation, continental drift and mid-ocean-ridge volcanism are driven by the transfer of heat from deep in the Earth to the surface by ascending plumes of heat and volatiles that create oceanic and continental hot spots (Niu 2020). This cooling and degassing process results in the release of heat, water vapour, CO_2 and other gases into the hydrosphere and atmosphere.

Because oceanic crust is far thinner than continental crust, most of this heat is released onto the ocean floor. This is also why most crustal extension takes place in the oceans. At present, the oceans occupy 70% of the surface of the planet, with 81% in the Southern Hemisphere covered by oceans and 61% in the Northern Hemisphere. The sites of oceanic crustal extension and hot spots are those where there is the highest heat flow (Figure 4.1).

The Sun is the Earth's primary source of surface heat and it is this heat exchange that drives weather and climate by the transfer of energy. The second largest source of heat is geothermal. The influence of this heat is not fully understood although there are suspicions that warm water 'blobs' and El Niño events derive from the cooling of submarine volcanics (Yim 2022).

It has long been established that seawater circulation in the top 5 km of the crust hydrates and alters sea-floor basaltic and ultramafic rocks by exothermic mineral reactions, thereby adding more heat to the oceans (German & von Damm 2003). For example, serpentinisation

Figure 4.1 Global heat flow distribution

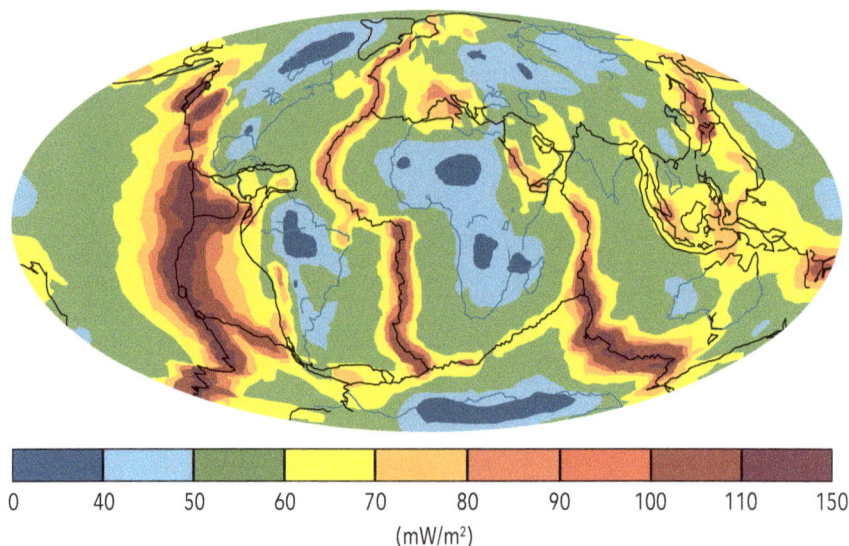

Source: Hamza and Viera 2012.

of 1 kg of peridotite underneath the basaltic oceanic crust raises the temperature of 1 litre of water by 50 °C. At any one time, between 1% and 2% of the volume of the oceans is circulating beneath the sea-floor. This circulation adds heat, removes magnesium, and adds calcium to seawater, which then combines with dissolved CO_2 to precipitate limestone (Joides Resolution 2011), thereby affecting the global CO_2 budget. Calcium is also added to the oceans by weathering and solution transport.

It can be seen, then, that the oceans are not a closed chemical or thermal system.

The low global average geothermal flow (91.6 mW/m²) is misleading because heat flow is concentrated around oceanic hot spots (for example, Hawaii, Galapagos, Iceland), mid-ocean ridges and their hydrothermal vent fields, and off-ridge guyots (Fuchs et al. 2021; Stål et al. 2022). At present, this heat is emitted via 30,000 geothermal and vent fields – there are 60,000 km of mid-ocean ridges, and at least 3.4 million off-axis guyots and seamounts in the Pacific Ocean and an unknown number in the other oceans.

Continental collisions that produce mountains, such as the Tibetan Plateau, change the weather and climate and affect the jet stream (Li et al. 2020). North of the Tibetan Plateau it is arid whereas to the south, India is affected by monsoonal rains.

The shape, position and size of continents affects ocean currents, too. This is because water has a high heat capacity; ocean currents carry low-latitude water to higher latitudes, which affects the weather and climate. All oceans have ridges, plateaux, seamounts and guyots, which divert deep-water ocean currents and affect upwelling. For example, the Indian Ocean has the Central Indian Ridge, the Southeast Indian Ridge, the Southwest Indian Ridge, Mascarene Plateau, Chagos–Laccadive Plateau, Carlsberg Ridge, Mozambique Plateau, Madagascar Plateau, Agulhas Plateau, Conrad Rise, Del Cano Rise, Kerguelen Plateau, Broken Ridge and Ninetyeast Ridge, all of which divert currents.

Large igneous provinces contain more than 100,000 km^3 of rock that was once molten. A large submarine igneous mass – such as the Central Atlantic Magmatic Province, the High Arctic Large Igneous Province or the Ontong Java Plateau – not only changes the topography of the ocean floor, thereby affecting heat transfer via ocean currents, but also adds monstrous amounts of heat, CO_2 and methane to the atmosphere, affecting the planet's carbon cycle (Deegan et al. 2023).

These can cause sea-level rises, climate changes, and the ozone layer to thin; mercury and other heavy metals are released into the atmosphere and there is short-term cooling due to the amount of sulphur dioxide in the air, which ends up as acid rain. As photosynthesis is reduced, plants die. Four of the big five major mass extinctions correlate with the emplacements of large igneous provinces (Grasby & Bond 2023).

The volcanoes of Iceland are part of the Central Atlantic Magmatic Province. The 1783–1784 AD eruption of Laki is a window into how sulphurous aerosols disrupt the thermal balance of the atmosphere and kill plants and animals (Thordarson & Self 2003). A supervolcano,[1] or a sustained series of eruptions, could change climate and create an

---

1   A volcano with a volcanic explosivity index of 8, which is the largest recorded value for this index.

extinction event. Furthermore, many of the Icelandic volcanoes are subglacial. The greatest climate change on Earth was in the Cryogenian period (720–635 Ma) and may have been initiated by a synchronous massive intrusion of igneous rock (Macdonald & Swanson-Hysell 2023).

Antarctica was once part of the giant supercontinent Pangea, which fragmented into two supercontinents Laurasia and Gondwana about 200 million years ago. About 100 million years ago, Gondwana fragmented into South America, South Africa, India, Australia, and Antarctica. The rifting of South America from Antarctica with the opening of the Drake Passage 34 million years ago established the Antarctic Circumpolar Current. This isolated Antarctica from the warmer south-flowing currents (Evangelinos et al. 2024), and so the Antarctic ice sheet started to form (Van den Ende et al. 2017). There are some 150 subglacial geothermal areas and volcanoes in active rifts as West Antarctica starts to fragment (Wei-Ping et al. 2009; Burton-Johnson et al. 2020; Day 2020). The polar position, and this isolation from warm water currents and subglacial heat, have a profound influence on the Antarctic ice sheet, making it one of the drivers of global climate.

It is well known that subaerial volcanic eruptions (that is, not underwater) change the weather and, if there is a sustained period of volcanism, can change the climate. Furthermore, after eruptions, there is reduced photosynthesis; and acid rain can form from the sulphurous aerosols. Terrestrial subaerial volcanoes rarely erupt lava and are explosive because of the high dissolved water content; volcanic eruptions are non-cyclical and can only be predicted a few days before an eruption. There are suspicions that planetary, solar, and lunar gravitational alignments may trigger increased seismicity, with resultant increased volcanic activity (Dumont et al. 2023). Sea-level falls reduce the hydrostatic load on ascending magma and induce submarine eruptions.

A large subaerial explosive volcano, like, for example, Krakatoa in 1883, can eject 30 km^3 of rock pulverised into dust, plus aerosols and gases (mainly steam). The explosive ejection of ash and aerosols – normally to an altitude of about 25 km – initially heats the air followed by cooling, resulting in the blockage of short-wave radiation, the formation of clouds, pressure changes, moisture redistribution, continental cooling, atmospheric circulation changes, and severe weather events. Such phenomena were recorded associated with eruptions of Tambora (10 April 1815), Krakatoa (26 August 1883), Tarawera (10 June 1886), Calbuco (7 January 1893), El Chichón (28 March 1982), Pinatubo (June 16 1991), Calbuco (22 April 2015), and Hunga Tonga (15 January 2022).

Satellite tracking of more recent eruption ash clouds (for example, El Chichón) showed the cloud circled the planet many times on a 21-day cycle, inducing a strong El Niño–Southern Oscillation (ENSO) (1982–1983), an extended wet season, and heavy rains with associated flooding and landslides (Rampino & Self 1984; Rampino et al. 1985). There are some 1,800 known terrestrial volcanoes, most of which are island arc and Andean, explosive, calc-alkaline volcanoes. Their explosivity is due to the large mass of dissolved water within the molten rock. Volcanic explosions result from the flashing and rapid expansion of dissolved water from silica-rich silicate melts. It was first noted that silicate melts are sensitive to their water content in 1798 (Spalloanzani 1798); later experimental petrology showed that water has a high solubility in calc-alkaline melts, proportional to the silica content (Tuttle & Bowen 1958), whereas CO_2 has a very low solubility.

The 16 June 1991 Pinatubo eruption blasted aerosols 55 km into the atmosphere. This eruption was followed by a global drought. The 15 January 2022 Hunga Tonga volcano ejected a thermal plume of ash, gas and pulverised rock 55 km into the atmosphere. About 146 megatonnes of water vapour entered the stratosphere, which comprises 10% of the stratospheric burden (Millán et al. 2023). This size of eruption could impact climate through surface warming caused by the radiative forcing from the excess stratospheric water. Regular rain 'bombs' years after eruptions are well known.

There was a rapid deglaciation in Antarctica and climate change in the Southern Hemisphere 17,700 years ago, coincidental with a 192-year-long series of halogen-rich volcanic eruptions from the West Antarctic's Mt Takahe volcano (McConnell et al. 2017). This led to large scale changes in atmospheric circulation and hydroclimate throughout the Southern Hemisphere. Ice core measurements show a spike in atmospheric acid 17,700 years ago, followed by a rapid increase in atmospheric CO_2, and a decrease in Southern Hemisphere dust.

Asteroid impacts and terrestrial supervolcano eruptions in the past – for example, at Yellowstone (USA), Taupo (NZ), Toba (Indonesia), and Kamchatka (Russia) – ejected from thousands to tens of thousands, of cubic kilometres of dust, and hundreds of million tonnes of sulphurous aerosols into the atmosphere. These can trigger sudden cooling of the Earth, followed by extinctions.

By contrast, submarine basaltic volcanoes have a lower silica content and reduced explosivity. They also contain a far higher concentration of dissolved CO_2 and a lower concentration of dissolved water (Dixon et al. 1995). If basalt melts contain a high potassium and sodium content – such as at Stromboli, Etna, Vesuvius and Erebus – they can dissolve large quantities of water and CO_2, and hence these are more explosive than other basaltic volcanoes (Allison et al. 2019).

The eruption of basalt at a temperature of 1,100 °C into basal seawater that is 2 °C results in a transfer of CO_2 and heat to the ocean, which in turn creates sea-surface temperature anomalies ('blobs'), pressure changes, circulation changes, atmospheric moisture redistribution, continental warming and severe weather events (Yim 2022). Examples of this scenario include the El Hierro volcano in the Canary Islands that erupted between October 2011 and March 2012 (Corrocedo et al. 2015; Somoza et al. 2017); the Axial Seamount (April-May 2015) (Wilcock et al. 2018); Wolf Island in the Galapagos (May 2015; January 2022); and Kilauea, Hawaii (July 2016–present).

Galactic climate cycles

Galactic climate cycles are roughly every 140 million years. Shaviv and Veizer (2003) theorised that the galactic cosmic ray flux is linked to climate variability. They suggested that at least 66% of the variance in oceanic palaeotemperature over the last 545 million years could be attributed to the galactic cosmic flux as the Solar System passes through the spiral arms of the galaxy. The integration of $\partial^{18}O$ data from calcite shells, the palaeogeographic distribution of marine ice-rafted debris, and terrestrial glacial debris, were used to plot icehouse and greenhouse conditions on Earth over the last 545 million years (Veizer et al. 2000). They noted that there is a poor correlation between modelled and calculated atmospheric CO_2 and climate over this period, suggesting that CO_2 models over this time need to be improved, or that atmospheric CO_2 might not be the principal driver of climate.

Veizer et al. (2000) showed that $\partial^{18}O$ was cyclical over 135 ± 9 million years. Palaeoclimate records and fingerprints of solar and cosmic ray activity using ^{10}Be, ^{14}C, and other cosmogenic isotopes show that extraterrestrial activity is responsible for climate variability on the scale of days to as long as millennia (Veizer et al. 2000; Lockwood 2007). Empirical observations indicate that there could be solar wind modulation of the galactic cosmic ray flux. The cosmic ray flux relates to low-altitude cloud cover on time scales ranging from days (Forbush phenomenon) to decades (sunspot cycles). It was postulated that a brighter Sun has an enhanced solar flux and solar wind, giving muted cosmic ray flux, fewer low-level clouds, lower albedo, and a warmer climate, while diminished solar activity results in a cooler climate (Svernsmark & Friis-Christainsen 1997; Svensmark 2000; Svensmark et al. 2017).

Cosmic ray flux variability arises from passages of the Solar System through the Milky Way on cycles of 143 ± 10 million years. The Shaviv and Veizer study, using astrophysics and geology, yields a consistent picture of ~140-million-year cycles of climate evolution on geological times scales. Lindzen (1997) showed that water is the key to climate. Negative feedback – such as increased cloud cover when the Sun has a reduced solar flux – implies that the water cycle is the thermostat of climate dynamics with water acting as both a positive (water vapour) and negative (clouds) feedback with the carbon cycle piggybacking on and modified by the water cycle (Lee & Veizer 2003).

The Sun has made about twenty cycles around the galaxy during the life of planet Earth and during this time the Sun has made many passages through the spiral arms of the disc. The six major ice ages that planet Earth has undergone are when the Earth was in Sagittarius–Carina (twice), Perseus, Norma, Scutum–Centaurus arm and Orion arm when there was increased cosmic radiation. Some, but not all, mass extinction events occurred at these times (Gies & Helsel 2005; Gillman & Erenler 2019).

On the basis of principal component analysis of the solar background magnetic field and changes in the Sun–Earth distance, Zharkova et al. (2023) suggest that Earth will be in a Grand Solar Minimum from 2020–2053, somewhat similar to the seventeenth century Maunder Minimum. This is discussed more fully later in the section on solar cycles, later in the chapter.

Orbital cycles

We have known about orbital climate cycles over much shorter time scales than tectonic and galactic cycles for more than a century. These position the Earth closer to or more distant from the Sun, hence the amount of energy received by the Earth is variable. Direct measurement of the Earth's orbit is integrated with data from ice cores, sediment cores from lakes and deep oceans, microorganism species, morphology and isotopic changes, and the distribution of glacial debris, to create a record of recent climate changes.

It is well known that the Milankovitch cycles are driven by changes in the shape of the Earth's orbit from nearly circular to slightly elliptical on a 100,000-year cycle. This is called eccentricity and has the planet closer to the Sun for about 10,000 years and more distant from the Sun for 90,000 years, producing warm interglacials and cold glaciations. During the many past interglacials, the rate of speciation increased, plants and animals migrated polewards, alpine tree lines rose in altitude, the area of vegetation and coral reefs increased, population of all animals increased and ice sheets retreated. The atmospheric CO_2 content increased as a result of thermally driven oceanic degassing. During glaciation, polar ice sheets, glaciers and sea ice expand; biodiversity, rainfall and crop yields decrease; dust storms and desertification increase; and plants and animals migrate towards the Equator.

During the penultimate interglacial, 128,000–116,000 years ago, warm climate animals such as water buffalo and elephants inhabited an ice-free Europe. Many of the heavy mineral sands of Australia were deposited during storm activity in this interglacial period, while sea levels and temperatures were higher than at present.

During the last glaciation, which came to an end only about 14,700 years ago, much of the Northern Hemisphere was covered by ice sheets,

Figure 4.2 Vostok ice core records showing 100,000-year eccentricity cycles

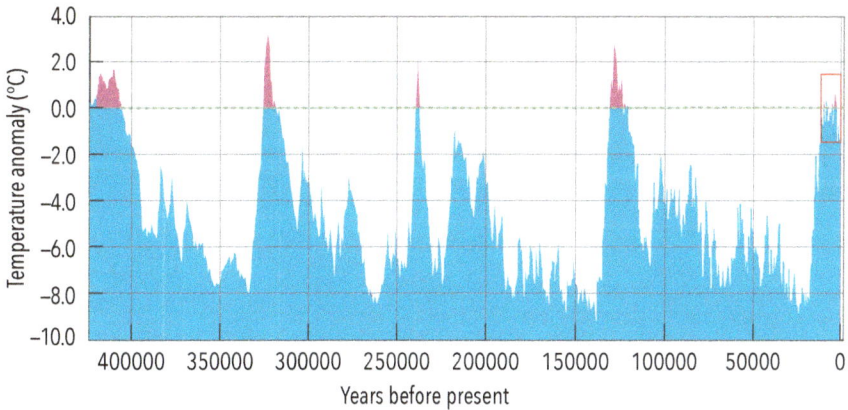

Vostok ice core records showing 100,000-year eccentricity cycles, interglacials and glaciations, wherein there are temperature spikes due to obliquity, precession and solar cycles. Note that the current interglacial is less intense and shorter than previous interglacials.

Source: Reconstruction of palaeotemperature from the Petit et al. (1999).

with Scandinavia covered by as much as 5 km of ice. Sea levels were 130 m lower than at present. Tasmania, the alps of Australia and New Zealand, and much of alpine South America and highland South Africa were covered by ice. Inland Australia, China, the United States of America and Africa had sandstorms and shifting sand dunes because they were drier than they are now. People walked from Papua New Guinea to Australia and on to Tasmania, forests retreated, plants and animals moved to lower latitudes, coral reefs died and there was a reduction in the global population of plants and animals. This was a time of accelerated extinction.

The current interglacial started 14,700 years ago with its peak 8,000–3,300 years ago when both temperature (+3 °C) and sea level (+2 m) were higher than they are today. The interglacial has peaked and a long-term cooling has commenced. The spikes of warming and cooling during this cooling period were probably driven by increased solar activity (Zharkova et al. 2023).

The tilt of the Earth's axis with respect to the Earth's orbital plane, mainly due to the gravitational pull of Jupiter and Saturn (roughly 41,000 years; obliquity), and the direction the Earth's axis of rotation, is

Figure 4.3 Greenland GISP ice core data showing solar-driven warmings and coolings

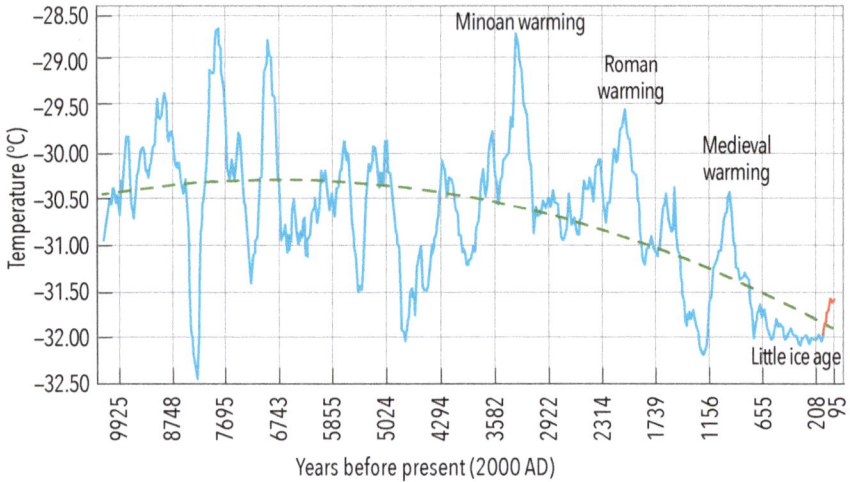

Greenland GISP ice core data showing the peak of the current interglacial (8,000–3,300 years ago) to the present, including solar-driven warmings and coolings.

Source: Greenland GISP ice core data from Alley (2000).

pointed varies between 21,000 and 29,000 years; referred to as precession. These factors caused climate changes that involved changes in ice and snow coverage, and hence changes in albedo – which further contributed to cooling by reflecting more solar energy back into space. At the same time evaporation from the sea formed ice, which was locked up in snow. This resulted in lower sea levels.

Sediment, ice core and microfossil studies show that between three and one million years ago, climate cycles matched the 41,000-year obliquity cycle. Muller & MacDonald (1997) show that over the last one million years, eccentricity cycles between 95,000 and 125,000 years dominated the climate record.

However, Milankovitch cycles cannot explain all aspects of climate over the last few million years.

Solar cycles

The Sun has a number of regular cycles and outbursts of energy, especially solar flares from sunspots. Solar activity can be cyclical and random and result in varying amounts of energy striking the Earth. The same occurs with other stars. These influence climate because they result in changes in the solar magnetic field that protects the Earth from solar particles and cosmic ray bombardment. The Sun has a larger effect in cooler times than in warmer times. There are variations in radiation emission, solar magnetic field intensity, solar magnetic polarity, particle emissions and solar surface convection. These changes affect the Earth in several ways that manifest as auroras, magnetic storms and changes in galactic cosmic rays, all of which produce climate change.

Changes in sunspots produce solar cycles of eleven years (also known as the Schwabe cycle), which is expressed as a 22.2-year Hale cycle.[2] These cycles have been known for 400 years. For example, Adam Smith[3] correlated sunspot numbers with grain prices. The same applies today. A low sunspot number leads to cold weather, low grain yields and high prices. The solar cycles are not perfectly regular and vary in length, shape, and intensity, or can even enter periods of an almost inactive state, called Grand Solar Minima, which occur roughly every 400 years (Usoskin et al. 2021). Other cycles – such as the solar wind ~87-year Gleißberg cycle (Feynman 1982); the ~210-year DeVries, or Suess, cycle (Lüdecke et al. 2015); the ~1,500-year Dansgaard-Oeschger cycle (Braun et al. 2005); and the ~2,400-year Hallstatt cycle (Usoskin et al. 2016) – all change climate.

The daily rotation of the Earth causes continual changes in the incoming solar energy and outgoing radiation. This gives us cool dawns and warm afternoons. The overlap of the daily solar cycles and the monthly lunar cycles gives us variations in short-term weather driven by air pressure and tides and currents in the oceans and atmosphere. The variation in the amount of incoming radiation caused by the tilting of the Earth's rotational axis gives us the seasons. This variation can be extreme near the poles.

These cycles are a surrogate measure of solar radiation and are shown on Earth as cycles of isotopic fingerprints that were initially generated

---

2   That is, a Hale cycle consists of two successive Schwabe cycles.
3   Adam Smith (1723–1790) was the founder of modern economics.

in the upper atmosphere by decreased solar activity. They include lake water levels, droughts and floods; sea-surface temperature and temperature-sensitive floating microorganisms; dust in ice sheets and ocean sediments; vegetation, pollen and spores from bogs and swamps; and historical records. The 22- and 1,500-year cycles dominate.

Solar minima and maxima are the two extremes of the Sun's eleven-year and 400-year cycles. During a Solar Maximum, the Sun has numerous sunspots, eruptions of solar flares, and ejections of billion-tonne electrified gas clouds into space. Sometimes several solar cycles exhibit greater activity for decades or centuries, such as during the Grand Solar Minima and Grand Solar Maxima. During Grand Solar Maxima, food production, population and wealth increases. Four solar cycles were out of phase in 2020. Solar physicists are suggesting that we are in a Grand Solar Minimum from 2020 to 2053, like the Maunder Minimum (1645–1715 AD), which was preceded by a lengthening of the Schwabe cycle (Miyahara et al. 2021). This is not unprecedented.

In the long and bitter Maunder Minimum (1645–1715 AD), the growing season was reduced, harvests failed, tree lines in the Alps dropped in altitude, ports were blocked by ice, there was mass famine and depopulation. The milder Dalton Minimum (1796–1820 AD) was similar to the Wolf and Spörer Minima.

The Wolf Minimum (1300–1320 AD) occurred immediately after the Medieval Warming (900–1280 AD). Rivers froze and there was crop failure (the Great Famine, 1310–1322 AD). The population, weakened by the famine, was greatly reduced in number by the plague (1347–1349 AD). The Spörer Minimum (1410–1540 AD) had the same effect as the Wolf Minimum (Loehle & Scafetta 2012).

Oceanic cycles

There are 60-year oceanic cycles that change ocean currents and sea levels. These have been known for thousands of years. Sexagenary cycle calendars and planting times – for example, the Chinese lunar calendar – are based on these cycles.

Sea levels, tides and polar sea ice are affected by lunar orbit cycles every 8.85 and 18.61 years – the 18.61-year cycle pushes warm water into the Arctic Ocean – at these times the Northwest Passage is navigable by ship. Sea levels can change by more than four metres.

The El Niño–La Niña events in the Pacific Ocean greatly influence rainfall and drought on both sides – in Australia and in South America, but less so in Europe. The combination of oceanic and solar cycles with El Niño–La Niña events can produce catastrophic weather with long droughts or huge floods. Surface waters in the Pacific Ocean are warmer during El Niño events, with mounting evidence that this is derived from sea-floor volcanicity.

Geology of carbon dioxide

All rocky planets in the Solar System contain a CO_2-bearing atmosphere. Carbon dioxide derives from degassing during planetary cooling, although some CO_2 on planet Earth derives from mineral reactions (for example, ocean degassing; thermal metamorphism $CaCO_3 + SiO_2 = CaSiO_3 + CO_2$), and from animal respiration.

The Earth's atmosphere is constantly evolving. The Archean[4] atmosphere had a similar nitrogen content to the modern one: the oxygen content was a millionth that of the modern atmosphere, whereas CO_2 was between ten and 2,500 times greater, and methane was hundreds to tens of thousands of times more than in the modern atmosphere (Catling & Zahnle 2020). Hydrogen and other gases were escaping into space, oceans were vapourised by asteroidal impacting, and early life started to add traces of oxygen and remove CO_2 from the atmosphere. Free oxygen of bacterial origin started to accumulate in the Proterozoic atmosphere that followed, which may have been due to supercontinent fragmentation (Eguchi et al. 2022). Despite a number of 'snowball Earth' events in the Proterozoic, where ice sheets are thought to have covered the Earth from pole to pole, the atmospheric CO_2 content was 20%. In the late Proterozoic there was a drawdown of CO_2 into dolomite reducing it to ~1% (Kaufman & Xiao 2003; Hoffman et al. 2017; Zhang et al. 2022). For context, today's atmosphere has

---

4   The Archean eon began about 4 billion years ago and continued for about 1.5 billion years.

only 0.04% CO_2. The Proterozoic atmosphere was also methane rich (Pavlov et al. 2003).

Every time continents are stitched back together, new mountain ranges form. Accelerated weathering of this new fresh rock draws down atmospheric CO_2 into soils. The erosion of these soils into sedimentary basins then sequesters CO_2 into sedimentary rocks (Bufe et al. 2021). There is no temporal relationship between periods of mountain building, basin formation, and global cooling suggesting that removal of CO_2 from the atmosphere does not drive cooling.

After the appearance of multicellular life, there was an explosion of other life forms and then predation. The drawdown of CO_2 into limestone continued. Marine life formed shells and reefs comprised of calcium carbonate, thereby removing CO_2 as well. Calcite fossils sequestered calcium carbonate from the atmosphere from the Cambrian explosion-of-life period until the present. The atmospheric CO_2 content has effectively

Figure 4.4 Plot of temperature and CO_2 over Phanerozoic time showing no temporal relationship between CO_2 and temperature

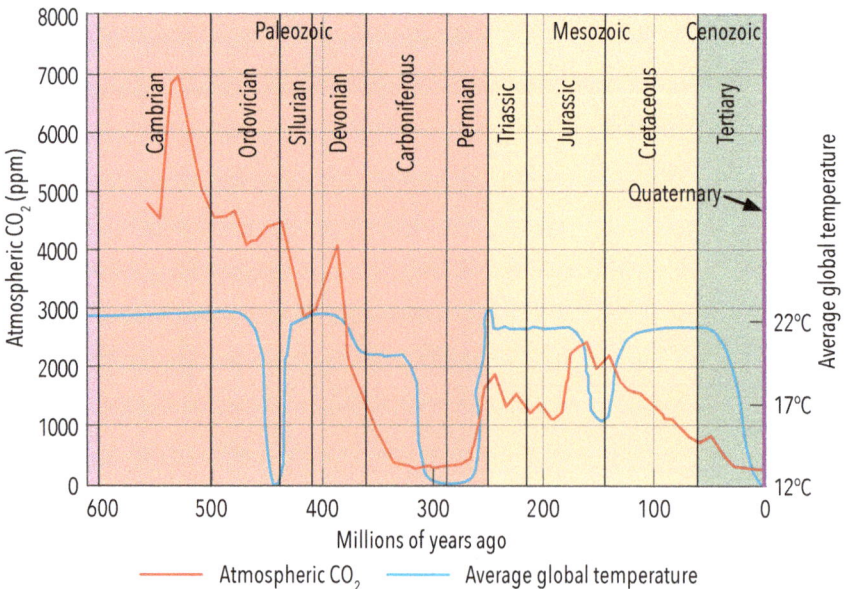

Source: Bernier and Kothavala 2001.

decreased from 0.7% to 0.04% with no relationship between temperature and atmospheric CO_2 over that time (Bernier & Kothavala 2001).

There has been long-term cooling of planet Earth over the last 50 million years (see Figure 4.4), with the current ice age starting 34 million years ago; furthermore, there is no valid calculation that might indicate when the current ice age will end. As pointed out in the introduction to this chapter, for more than 80% of geological time there was no ice on Earth, the planet was warmer and wetter, and sea levels were higher.

Warm CO_2 has long been pumped into glasshouses as plant food to stimulate growth and reduce water consumption. C_4 plants[5] evolved in response to the decrease in atmospheric CO_2 to 0.05% between 35 and 24 million years ago (Sage 2003); however, if atmospheric CO_2 is greatly reduced further, then plant life on Earth will die (Gerhart & Ward 2010). The next extinction of life may, in fact, be due to a reduction of atmospheric CO_2.

Conclusions

The history of the planet shows that past climates were not driven by atmospheric CO_2 but were driven by an interplay of orbital, solar, galactic and tectonic processes. Random events, such as an asteroid impact, can also change climate. Climate models ignore basic planetary processes, so unsurprisingly, even over time periods of mere decades, they have been shown to be incommensurate with the actual measurements.

It has not been shown that human emissions of the gas of life, CO_2, drive global warming. If this were shown, then it would also have to be shown that the natural CO_2 emissions (Σ97%) do not drive global warming. This also has not been done. The climate history over billions of years, the Phanerozoic history of temperature and CO_2 over hundreds of millions of years, ice core data over thousands of years, proxies over hundreds of years and modern measurements over 150 years all show that there is no relationship between temperature and the atmospheric CO_2 content.

---

5   Plants that have evolved to cycle CO_2 into four carbon sugar compounds using a photo-synthetic process that includes a particular enzyme.

5 Why Climate Models Fail

Christopher Monckton of Brenchley

At the Russian Cities' international conference on climate change in Moscow in 2017, Dr V.A. Semenov, deputy director of the Russian Academy of Sciences, led a high-level discussion on why, after decades of climate modelling, it had proven impossible to constrain 'equilibrium doubled-carbon dioxide (CO_2) sensitivity' (ECS). In this chapter, Semenov's question will be addressed by showing that the general circulation models (GCMs) of climate are incapable of constraining the breadth of the long-predicted 3 [2 to 5] Kelvin (K) interval of ECS.[1] Through a fundamental error of physics, the predictions based on these models' outputs are now proving to be excessive. The error aggravates the defects of the integrated-assessment models (IAMs), which, in turn, are used by economists to justify the costly measures being put in place in order to abate anthropogenic emissions. After correction, such measures are probably unnecessary and certainly not cost-effective.

GCMs predict that an anthropogenic forcing equivalent to doubling the CO_2 concentration in the air will drive 1.2 K of direct warming (Hansen et al. 1984), also known as 'reference doubled-CO_2 sensitivity' (RCS). However, feedback processes, notably the increase in the capacity of the air to carry more water vapour as it is directly warmed, are known to amplify that direct warming. The ECS is the sum of the RCS and

---

1   In climate, calculations are in degrees absolute, known as Kelvin. Zero degrees Celsius, the freezing point of water, is equal to 273 Kelvin, but a change in temperature is the same in Kelvin as in degrees Celsius.

this feedback response. Since the late 1970s, GCMs have predicted an ECS in the order of 3 [2 to 5] K (for example, IPCC 2021), implying a 1.8 [0.8 to 3.8] K feedback response. Since the uncertainty in the 1.2 K RCS is thought to be as little as ±10%, feedback response in the order of 1.8 [0.8, 3.8] K (currently thought to contribute as much as 60% [40% to 75%] to ECS) is the chief source of the ±70% uncertainty in the ECS.

It will be shown that, by a fundamental error of control-theoretic physics in the diagnosis of feedback strengths from models' outputs, the feedback response, and consequently ECS itself, are overstated. After correction, the climate emergency ceases to exist. It will also be shown that, even if all nations achieved net zero emissions by 2050, only 0.1 to 0.2 K global warming would be prevented by that year, at a cost amounting to quadrillions of dollars.

The world is not warming as fast as models predict

Actual anthropogenic emissions (Friedlingstein et al. 2022) are closer to scenario A (IPCC 1990) than to scenarios in later IPCC reports. However, the 0.3 K/decade scenario-A, midrange, medium-term prediction is still double the observations. The most recent 3 [2 to 5] K predicted ECS (IPCC 2021) is equivalent to ten decades of predicted 0.3 [0.2 to 0.5] K/decade medium-term warming (IPCC 1990), while observed warming in the 33 years from 1990 to 2023 was only 0.14 to 0.2 K/decade (UAH; HadCRUT5).

Defects in models

Climatologists assume that:

- GCMs, although far from perfect, are the best available way of predicting global warming.
- although weather can change unpredictably in the short-term, longer-term climatic trends (longer than 30 years) *are* predictable.

However, models' excessive global-warming predictions in the 33 years since 1990 raise legitimate questions about these assumptions. Why do the models run so much hotter than the climate? And can they tell us anything about future warming?

Not the least of the defects in GCMs is their misrepresentation of how the air warms at altitude in the tropics. The GCMs predict that the rate of global warming will not be constant at all altitudes in the tropics, but will be approximately twice as rapid at 9 km above ground level than at the surface (Hansen et al. 1984; IPCC 2007, fig. 9.1; Christy et al. 2018), when the reality is that it has declined (ESRL 2023; Karl et al. 2006) – the opposite of what the GCMs predict. One possible reason is subsidence drying (Paltridge et al. 2009). Another is that models make no allowance for earlier tropical afternoon convection as the tropical surface warms – one of many processes that models do not yet represent.

McKitrick & Christy (2020) showed that the observed rate of upper-air warming was 0.17 K/decade, while the average prediction in 39 current GCMs was 2.4 times greater, at 4 K/decade. This over-prediction by more than double is why models predict a large amplification by water vapour feedback and thus large ECS. At midrange, all feedbacks other than the water vapour feedback broadly self-cancel (IPCC 2013).

Another error of feedback evaluation lies in GCMs' treatment of net cloud feedback as strongly positive. The IPCC (2021) assesses cloud feedback strength as +0.42 [–0.1 to +0.94] W/m²/K, again, a broad and poorly constrained uncertainty interval. However, cloud feedback is more likely to be negative than positive; it is the sum of the negative cloud albedo feedback, or parasol effect, by which clouds shade the surface from the Sun during the day and reflect its light harmlessly spaceward, and the positive blanket effect, by which clouds inhibit heat loss from the surface at night. The parasol effect is predominant, as can be seen by looking at an image of the Earth from space through half-closed eyelids. The reflectiveness of the clouds is striking.

By another error, GCMs fail to conserve either mass or energy – a requirement essential in any physics-based model. The change in the global energy budget arising from increases in anthropogenic forcing represents only 1% of average inbound and outbound radiation. Irving et al. (2021) found that models fail to conserve energy in the ocean and at the top of the atmosphere, and also fail to represent the atmospheric water-vapour budget correctly.

One reason for this error is the drift caused by the propagation of uncertainty. Models start with initial data that are uncertain and calculate the evolution of climate by numerous small time-steps over decades or centuries. At each step, the uncertainty grows. Frank (2019) demonstrated that propagation of the annually averaged ±4 W/m² published uncertainty in just one climate variable – the long-wave cloud forcing – exceeds the entire 0.04 W/m² annual mean anthropogenic greenhouse-gas forcing (Butler & Montzka 2021) 100-fold. Once that ±4 W/m² uncertainty is propagated through the time-steps over a century, any warming predictions between −15 K and +15 K (including all current predictions) are, statistically speaking, mere guesswork.

Even where measurements are correct, GCMs do not always take them sufficiently into account. For instance, the warming of the late 20th century coincided with a period of less cloud that let more sunlight reach the surface. This global brightening contributed more to the warming of that period than we did. Pinker et al. (2005) reported an increase in solar radiation reaching the surface at an annual average rate of 0.16 W/m² from 1983 to 2001. Yet net anthropogenic forcing occurs at less than a fifth of that annual rate (NOAA 2023). Global brightening also accompanied the warm Northern Hemisphere summer of 2023.

Modellers often seek to validate their predictions of global temperature change by studying the early climate. In the IPCC's First Assessment Report (1990, Fig. 7.3c), Hubert Lamb's graph of the past 1000 years' global temperature change shows the Mediaeval Warm Period (MWP) as warmer than 1990 and the Little Ice Age (LIA) as cooler. However, IPCC (2001) published a graph derived from breadths of annual growth-rings in 1000-year-old trees, contradicting the earlier understanding by eradicating the MWP and LIA. It showed stable temperatures until 1900, followed by a sharp 20th-century increase, suggesting that current temperatures are unprecedented in 1000 years. The IPCC (2021) shows a similar 1000-year temperature graph.

Lamb's graph is nearer the truth than the IPCC's current graph showing temperature change during the past 1000 years. Since sea-level changes are correlated with temperature changes, comparing the competing graphs of millennial temperature with a graph of millennial sea-level evolution (for example, Grinsted et al. 2009) shows the sea level was higher than today during the MWP and lower in the LIA, matching Lamb's graph, but not IPCC's current graph.

Worse, near-undetectable tuning of models to achieve a politically desired result is too easy. Thanks to tuning, models are made to represent past temperature changes correctly. However, they diverge markedly from one another in their predictions. Mathematically, the climate system behaves as a chaotic object; minuscule changes in initial conditions may lead to unpredictably large changes in outputs (Lorenz 1963). A chaotic object always changes for a reason, but we cannot predict changes accurately unless we have an unattainably exact knowledge of the conditions at the chosen starting point. In climate, most initial conditions are unknowable to anything like a sufficient precision. Not the least of these insufficiently well-resolved initial conditions is the net feedback strength.

The temperature-feedback error

Climate modellers do not directly use feedback mathematics. Instead, attempts (for example, Bony et al. 2006; Vial et al. 2013) are made to diagnose feedback strengths from GCMs' outputs.

Feedback response is an additional warming driven by and proportional to the direct or reference temperature (before adding feedback response) at a given moment. In climate, that moment is when the climate resettles to equilibrium following a forcing equivalent to doubling the CO_2 in the air compared with the pre-industrial value in 1850.

Feedback strength is denominated in Watts per square metre per Kelvin of the reference temperature at a given moment. A positive feedback response, in Kelvin, amplifies temperature, while a negative feedback response diminishes it. The sum of reference temperature and its feedback response is the equilibrium temperature after restoration of equilibrium following a disturbance such as doubling atmospheric CO_2.

In pre-industrial 1850, the climate was at equilibrium; there would be no trend in global temperature for 80 years (HadCRUT5). At that time, equilibrium temperature was 287.5 K. The reference temperature was 267.5 K – which is the sum of the 259.6 K surface emission temperature that would prevail without greenhouse gases and the 7.9 K direct warming by the natural greenhouse gases then present (Mein-shausen et al. 2017; IPCC 2007, table 6.2). The feedback strength was the 20 K difference between the equilibrium and reference tempera-tures. The reference signal relevant to derivation of ECS thus becomes the 268.7 K sum of the 267.5 K reference temperature in 1850 and the 1.2 K RCS.

Climatologists borrowed feedback mathematics from control theory in engineering physics in 1984. Due to increasing specialisation in the sciences, they did not realise that feedbacks respond not merely to a change in reference temperature, such as the 1.2 K RCS, but rather to the entire 268.7 K reference temperature itself. Thus, 96.5% of total feedback response is to the 259.6 K emission temperature; 3% is to the 7.9 K natural reference sensitivity; and only 0.5% is to the 1.2 K RCS.

The system-gain factor A is the ratio of equilibrium temperature E to the 268.7 K reference temperature R. Moreover, A is equal to $(1 - \Lambda / P)^{-1}$, where Λ is the +2.06 W/m²/K midrange feedback strength (IPCC 2021), while midrange P is the 3.22 W/m²/K Planck sensitivity parameter (ibid.). Thus, climatologists' midrange A would be 2.776. The impossible implication is that, after a doubling of CO_2, global temper-ature, E (the product of R and A) would be 746 K, which is twice the boiling point of water.

In 1850, the true system-gain factor A was the ratio 1.075 of the 287.5 equilibrium surface temperature to the 267.5 K reference temper-ature that year. Since the 1.2 K RCS increases that reference temperature by only 0.44%, there is no good reason to expect any change in A upon doubling CO_2 compared with 1850. Assuming no change in A, equilib-rium temperature E would be the 288.8 K product of A and the 268.7 K reference temperature R. Then ECS (simply E – R) would be 1.3 K – well below the 2 K lower bound of IPCC's ECS interval.

The central question in climate-sensitivity studies thus reveals itself. In response to the 0.44% increase in reference temperature represented by RCS, do the net feedback strength Λ and the system-gain factor A increase? Since feedbacks respond to the whole reference temperature R, even an infinitesimal change in A would drive a large change in E, and hence in ECS, compared with the 1.3 K using the value of A in 1850. Climatologists have not hitherto recognised this question, since they incorrectly imagine that the system-gain factor is the ratio not of equilibrium to reference temperatures (of E to R) but of equilibrium to reference sensitivities (of ECS to RCS).

In reality, uncertainties in true feedback strengths are so large that feedback analysis cannot be used to predict global warming. Since feed-back strength Λ is equal to P (1 – R / E), the IPCC's 3 [2 to 5] K ECS implies Λ on 0.24 [0.23, 0.26] W/m²/K. Thus, an increase of only 0.01 W/m²/K would increase ECS by 1 K. Yet we cannot know feedback strength to any such precision. Feedback analysis is thus valueless for temperature prediction; yet IPCC (2021) mentions 'feedback' more than 2500 times.

Comparing the curves of ECS against corrected and uncorrected feedback strengths shows how hypersensitive ECS is to tiny increases in true feedback strength (see Figure 5.1). Just 0.03 W/m²/K separates 5 K from 2 K ECS.

ECS is thus far more sensitive to changes in feedback strength than climatologists appreciate. In effect, they have forgotten to allow for the

Figure 5.1 Hypersensitivity of ECS to increases in true and variant feedback strengths

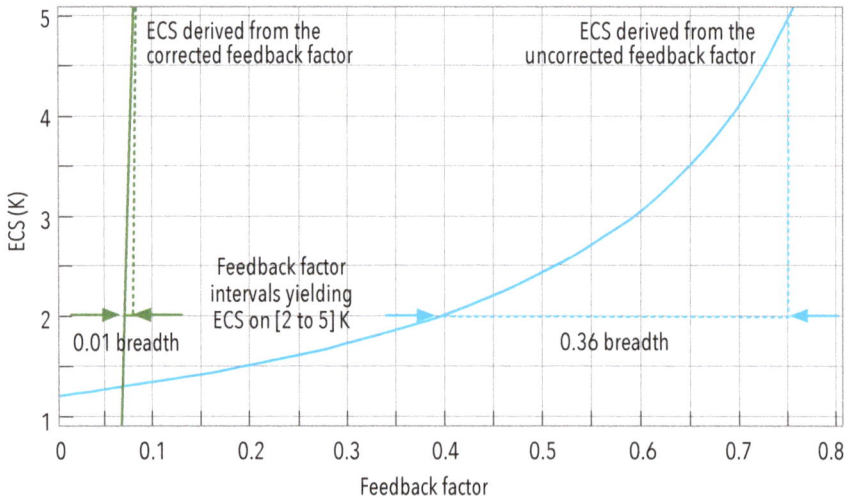

Source: Christopher Monckton of Brenchley.

fact that the Sun is shining. After correction of climatologists' error, large and potentially dangerous warming becomes merely one possible outcome, and it is not the expected one.

There are several reasons, independent of GCMs' outputs, to suspect that anthropogenic global warming will prove small enough and slow enough to be harmless and even net beneficial. First, since the climate system is powerfully thermostatic, there is little reason to suppose that A will have changed at all in the industrial era, in which event ECS is only 1.3 K and only 1 K further global warming may be expected this century – well within natural climatic variability. Second, midrange warming predicted by the IPCC (1990) was 0.3 K decade^{-1}, implying 3 K ECS, but only 0.14 K/decade is observed, implying ECS of order 1.4 K, and again little more than 1 K further warming by 2100. Third, energy-budget analysis suggests ECS on 1.3 [0.9, 2.0] K. The coherence of these three independent derivations of ECS, each simpler and less error-prone than diagnosing feedback strengths from GCMs' outputs, suggests that global warming may not, after all, be an emergency.

Defects in integrated-assessment models

Interdisciplinary compartmentalisation bedevils economists' attempts to compare the benefits and costs of attempting to mitigate future warming with those of adapting to any of its consequences that are net adverse. When climatologists' predictions are excessive, economists assume they are correct. But even if they were correct, there are strong economic reasons not to take climate action.

First, some 70% of new greenhouse gas emissions arise in nations that are exempt from the Paris Climate Accords (British Petroleum (BP) 2019). Emissions-abatement legislation in the chiefly Western nations selectively targeted by the Accords has greatly increased their electricity and compliance costs, and has set their terms-of-trade at a profound, and deepening, disadvantage against other nations that are exempt. Electricity prices (globalpetrolprices.com 2023) in Germany, Denmark and Italy, at $0.80 kWh^{-1} for households and $0.60 kWh^{-1} for businesses, and at $0.41 and $0.43 kWh^{-1} respectively in the UK, exceed the $0.10 and $0.08 kWh^{-1} in India and China sixfold to eightfold.

Russia has virtually no renewable generation. China (France24 2022), India (Reuters 2022) and Pakistan (Reuters 2023), with more than one-third of the global population between them, are greatly expanding their coal-fired generating capacity, not least to accommodate production that has been priced out of Paris-obligated nations by the increasingly costly emissions-abatement measures, the chief cause of the rapidly growing disparity between Western and Eastern electricity prices. Particularly where displaced manufactures are energy-intensive, eastward transfer of Western jobs and industries increases global emissions (the opposite of what was intended), since manufacturing in Paris-exempt nations emits more CO_2 per unit of production than in the West.

Accordingly, despite decades of gradually increasing economic self-destruction in the West, the uptrend in annual greenhouse-gas forcing has remained very close to linear since 1990 (see Figure 5.2). The West's self-harming attempts at mitigation have made no discernible difference to that uptrend.

Figure 5.2 Greenhouse-gas forcing since 1990

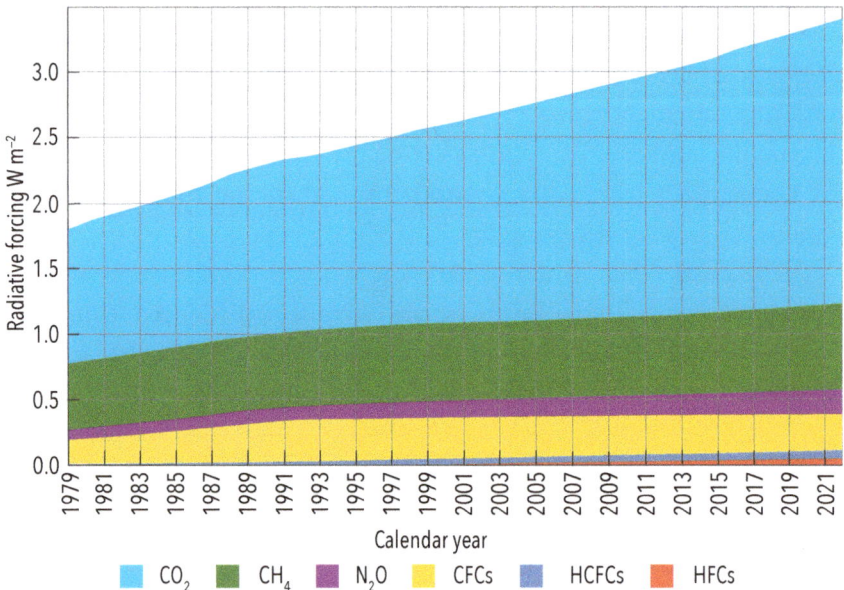

Source: NOAA AGGI 2023.

Statistical analysis based on published intervals of doubled-CO_2 effective forcing (IPCC 2021, p. 925), transient doubled-CO_2 climate response TCR thereto (ibid, p. 93) and an estimated interval of emissions abated if all nations moved straight from current emissions to net zero in 2050, shows that, despite the quadrillions that would be spent on attaining net zero, only 0.2 [0.1 to 0.3] K global warming would be abated even if global net zero were attained by that year. After correction to allow for the factor-2 excess of originally-predicted compared with subsequently-observed warming, the midrange warming forestalled by global net zero would be less than 0.1 K by 2050.

The cost would be disproportionate. McKinsey (Kumra 2022) estimated the capex cost alone of worldwide net zero as $275 trillion – half of global corporate profits. Since opex is typically at least twice capex, total cost might be $1 quadrillion. Then every $1 billion spent would prevent one five-millionth of a degree of warming if GCMs' midrange predictions were sound, and one ten-millionth if not, or one 16-millionth based on global extrapolation from the UK grid authority's published $3.6 trillion cost of net-zeroing the UK grid (National Grid ESO 2020). The grid contributes only a quarter of national emissions (Office for National Statistics (ONS) 2021). Since the UK emits less than 1% of all emissions, global net zero might cost $1.5 quadrillion. The United States, which represents 15% of global emissions (Environmental Protection Agency (EPA) 2023), would on its own reduce global warming by less than 0.015 K even if it attained net zero.

Conclusion

The extent to which the notion of climate emergency is founded upon GCMs' outputs is not widely appreciated. Their defects make them valueless for predicting the rate of global warming. Two such defects – failure to account for propagation of uncertainty and tenfold overstatement of feedback strength – irremediably prevent models from making predictions that are any better than guesswork.

After correction of the long-standing and pervasive control-theoretic error described here, the reduced probability of large warming swings the risk-reward ratio against climate action. Any mitigation that is inexpensive enough to be affordable will be ineffective, while mitigation expensive enough to be effective will be as unaffordable as it is unachievable – and probably unnecessary. Adaptation, to the limited extent necessary, is the rational economic choice. For energy security and affordability while coal, oil and gas reserves endure, thermal generation may, after all, safely be retained. The planet will come to little harm thereby.

6 Global Climate Lysenkoism: The Politics of Preferring Models Over Observations

Emeritus Professor Aynsley Kellow

> Since all models are wrong the scientist cannot obtain a 'correct' one by excessive elaboration. On the contrary following William of Occam he should seek an economical description of natural phenomena. Just as the ability to devise simple but evocative models is the signature of the great scientist so overelaboration and overparameterization is often the mark of mediocrity.
>
> George Box (1976: 792)

The making of public policy is a difficult, complex process. Bismarck once remarked that politics was the art of the possible, although in 1969 Daniel P. Moynihan perhaps better summed up the contemporary state of the art of government as the art of the *impossible.*

Policies have hypothetical status. They are devised so that if certain actions are taken, some desired outcome will result. They can succeed or fail as instruments, but they can also fail because things have gone awry during implementation – or they can go awry in terms of norma-tive justification (Kerr 1976). But ultimately, they must be based upon accurate knowledge of causes and effects; the policy process (much like the scientific process) is essentially one of learning – the detection and correction of error – because both are prone to error.

Policy making involves the exercise of power, with participants bringing different perspectives to the process depending on their inter-ests and their normative beliefs. As Miles's Law puts it, 'Where you stand depends on where you sit' (Miles 1978). We rely on the political process to reach decisions on complex matters. While politics is messy, it (albeit imperfectly) ensures that elements necessary to the policy – both relating to norms, or facts and theories – are not neglected. They might not prevail, but at least they are not ignored.

The complex mix of causative elements, including norms and the implications of proposed measures is why, as Roger Pielke Jr (2007) has pointed out, there is no linear relationship between science and policy. Any particular piece of scientific information does not lead to any particular policy. Even if we accept someone's views on climate science, that does not mean that we should choose mitigation over adaptation. And the views of expert climate scientists are not all that matters. Experts, after all, know much about their field of expertise but remain ignorant about a multitude of other fields. For this reason, a common aphorism holds that experts should be on tap, but not on top.

Some climate scientists seem unaware of that aphorism. Five lead authors of IPCC reports told *The Guardian* in December 2023 that scientists should be given the right to determine policy prescriptions and possibly even to oversee their implementation by the parties to the Framework Convention on Climate Change. Besides revealing a remarkable naivety about the global policy process, this also revealed a remarkable arrogance on the part of the scientists – that they would claim the power to determine how what they saw as an urgent need to mitigate greenhouse gas (GHG) emissions might be balanced against economics, social factors and norms prevailing among the 195 parties to the United Nations Framework Convention on Climate Change (UNFCCC).

The desideratum in policymaking is therefore 'evidence-based policy' (Head 2013), but what constitutes evidence? And how can we ensure that the best evidence is brought to bear? The fact that scientific findings can be wrong, even in medical science (Ioannidis 2005), makes it difficult to have confidence in the findings of science on something as complex as the global coupled ocean-atmosphere.[1] Karl Popper recognised the inherent fallibility in all scientific enquiry, and advocated that science should be

---

1   The 'global coupled ocean-atmosphere' refers to the integrated system where the Earth's oceans and atmosphere interact with each other. This interaction is complex and involves the exchange of energy, moisture, and momentum.

judged by how well its predictions had withstood repeated attempts at falsification. We can have greater confidence in those theories that have withstood repeated attempts at falsification but should expect that they might still prove to be wrong.

For this reason, science is best seen as a *process* (Hull 2019) and contests between alternative perspectives should be encouraged – just as liberal democracies, with a lively contest of ideas and preferences, yield better decisions than autocracies. Unfortunately, the complexity of climate science was deemed to pose a barrier to policymakers being able to respond 'appropriately' to the risks of climate change and consequently an *institution* was created in the form of the Intergovernmental Panel on Climate Change to provide them with something that has since come to be regarded as 'The Science' – the definitive word on the causes and effects of climate change and the best way to respond to the risks.

Adaptation has been eschewed from the beginning of the 'climate alarm' period, while the pursuit of mitigation has led to the corruption of climate science. At first, this was noble cause corruption (Kellow 2007), but mitigation-oriented policy responses created their own interests, which have joined the norms-based actors in a classic 'Bootlegger and Baptist coalition' (Yandle 1983). There is now more traditional corruption, which supports the production and reporting of science that, in turn, supports their interests.

Groupthink has seen the emergence of a scientific consensus of 'The Science' that is fundamentally anti-science. Groupthink is a dysfunctional phenomenon within groups that puts the emphasis on harmony and conformity; it minimises conflict, and emphasises consensus decision making at the expense of critical evaluation (Janis 1972). The group usually overrates its own abilities and underrates those of its opponents, often dehumanising them. Science is a *process* that encourages diversity and the production of evidence, not a destination. As Popper put it, knowledge advances through disagreement. Climate science is now dominated not by observational evidence, but by 'projections' from models that are treated as evidence.

Modelling as the basis for public policy did not fare too well during the Covid-19 pandemic, when alarmist modelling by people such as

Professor Neil Ferguson (Imperial College London) led to draconian responses and drew attention to his earlier alarmist modelling on bovine spongiform encephalopathy (BSE), which became the basis for some very costly policy responses. Climate modelling has generated projections that have proven to be both exaggerated and wrong.

Models should not be tested against the data from which they are constructed, but against the accuracy of their projections. Much of the 'evidence' upon which climate policy is based is not observational data, but projections produced from models, and therein lies a massive problem. As statistician George Box wrote in 1976, 'All models are wrong, some are useful' (Box 1976, p. 792). To accept model outputs as the definitive truth, as 'The Science', is extremely problematic, yet that is where we are basing policy on – model projections in scholarly articles that inevitably include the word 'could' in their abstracts. Popper would say that these are conjectures without any attempt at refutation.

Moreover, not only is the evidence for action based on models, but the targets for policy responses are also set in terms of projected mean global temperatures (MGTs) arising from changes in atmospheric concentrations of carbon dioxide (CO_2) resulting from reductions in emissions. In chapter 8, William Kininmonth shows the problems in considering MGT as a relevant metric, because temperature changes vary widely by region. For example, looking at the maximum temperature observations for the period from 1887 to 2013 in Southeast Australia, it suggests there has, in fact, been more than 1.5 °C of cooling to 1960 and more than 1.5 °C of warming since then (Marohasy & Abbot 2016). The rationale for action and the design of policy responses are therefore *both* predicated on models, so it is vitally important that, while inevitably wrong (as Box would note), they are as accurate as possible.

Climate models have been elaborated and adjusted to better replicate past observations, but the test of the accuracy of any model is the accuracy of the predictions it generates, rather than whether it can replicate the data used to produce it. Babyak (2004, p. 414) holds that if a sample is used to construct a model, or if you want to choose a hypothesis to test, a rigorous scientific test of the model or the hypothesis cannot be made using that same sample data. Freedman, (1991, pp. 306–307) argues that replication and prediction of new results provide a harsher and more useful validation regime than statistical testing of many models on one data set, which is the approach followed by the IPCC (see also Frické 2015).

Moreover, the work of Marohasy (2020) has shown that historical temperature records have been adjusted ('homogenised') and not always validated when changes have been made from the mercury thermometers that have been used historically to modern digital probes. Similarly, the Surface Stations Project in the United States revealed a lack of quality control in the siting of many temperature recording sites (Fall et al. 2011). Those that were poorly sited (near asphalt car parks and air conditioner plants) showed more warming, confirming that the United States Historical Climatology Network was contaminated by factors collectively referred to as the 'Urban Heat Island' effect (UHI).

Therefore, a crucial question is: how well have climate models performed in their predictive capabilities. The short answer is, not very well.

In March 2017, John Christy (2016, p. 9) presented evidence in United States Congressional testimony to the House Committee on Science, Space and Technology on 'Climate Science: Assumptions, Policy Implications and the Scientific Method' showing that the Fifth Assessment Report of the IPCC indicated that the observational record was more consistent with model runs with no GHGs than with those with GHGs. Christy also showed how a key prediction of the theory of Catastrophic Anthropogenic Global Warming – the presence of a tropical tropospheric 'hot spot' – was absent, thus falsifying a key prediction (Christy 2016, p. 2–4) (see Figure 6.1). Nicola Scafetta (2022; 2023; 2023a) has reported a similar warming bias in climate models.

Christy's evidence took on even greater significance when van Wijngaarden and Happer (2019) examined the measured (rather than modelled) changes in the loss of radiation from the atmosphere and found that the net forcing increase from methane (CH_4) and CO_2 was about 0.05 watts per square metre per annum, which, they estimated, with other things being equal, would cause a temperature increase of about 0.012 °C annually, or 0.12 °C per decade – consistent with the increase reported by Christy and his colleague Roy Spencer at University of Alabama, Huntsville (UAH) – over the oceans (free from any urban

Figure 6.1 Five-year averaged values of annual mean (1979–2015) global bulk temperature

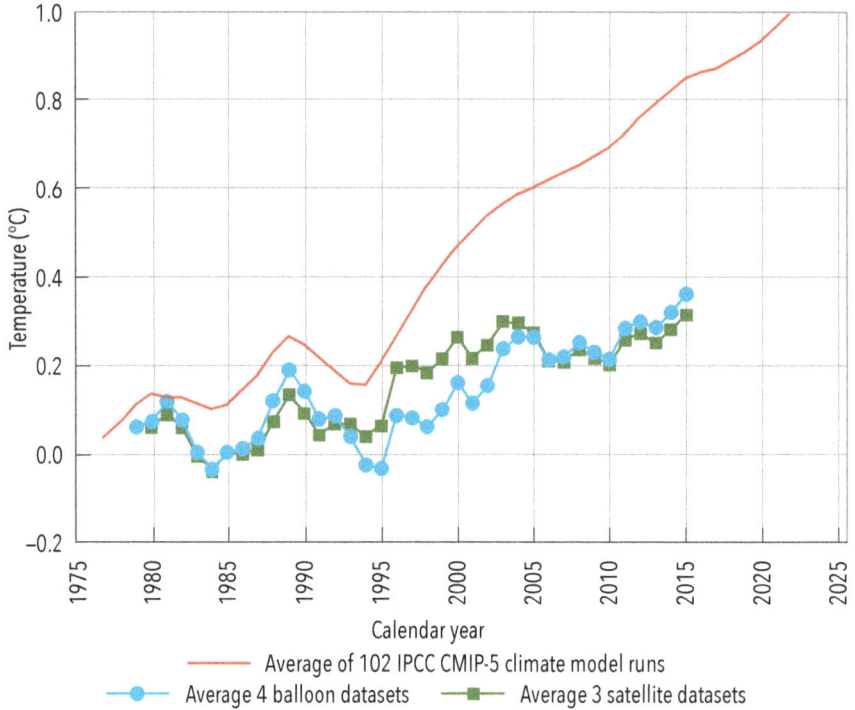

Source: JR Christy, University of Alabama in Huntsville Model output – KNMI Climate Explorer. Five-year averaged values of annual mean (1979–2015) global bulk (termed 'mid-tropospheric' or 'MT') temperature as depicted by the average of 102 IPCC CMIP5 climate models (red), the average of three satellite datasets (green – UAH, RSS, NOAA) and four balloon datasets (blue – NOAA, UKMet, RICH, RAOBCORE).

heat island effect) derived from more than 40 years of global satellite measurements. A recent uptick has seen the UAH trend to 0.14 °C per decade, possibly as a result of the Hunga Tonga eruption in January 2022, which injected into the atmosphere an extra 13% of water vapour, which is overwhelmingly the most powerful GHG (Khaykin et al. 2022; Sellitto et al. 2022; Jucker et al. 2023).

The HITRAN (High Resolution Transmission) molecular spectroscopic database allows the simulation and analysis of the transmission and emission of light in gaseous media, especially planetary atmospheres. This database allows the absorption of Earth radiation at its current temperature of 288 K to be determined for each individual constituent of the atmosphere, and for the combined absorption of the atmosphere as a whole that ensures the Earth is 33 K warmer than the 255 K. Using this data, Coe, Fabinski and Wiegleb (2021) concluded that water vapour is responsible for 29.4 K of the 33 K warming; CO_2 contributes 3.3 K; and CH_4 and nitrous oxide (N_2O) combined contribute just 0.3 K.

Climate sensitivity to future increases in CO_2 concentration was calculated to be 0.50 K, including positive feedback effects of H_2O, while climate sensitivities to CH_4 and N_2O were almost undetectable at 0.06 K and 0.08 K respectively.

Coe, Fabinski and Wiegleb strongly suggest that increasing levels of CO_2 will not lead to significant increases in the Earth's temperature, which is consistent with the calculations of Arrhenius (1896)[2] more than a century ago and helps explain why observations have tracked below model projections, which incorporate higher assumptions of climate sensitivity. They also found that increases in CH_4 and N_2O will have very little discernible impact. When these are combined with findings that recent increases in water vapour have not resulted in amplifying the warming (Paltridge et al. 2009), they provide strong reasons for questioning the projections obtained from models. This is especially so when those models are fed highly unrealistic emissions scenarios – like RCP8.5 (Pielke & Ritchie 2021) – and when these models are used to set emission mitigation targets as if we can 'dial in' a desired temperature outcome.

Van Wijngaarden and Happer (2020) reached similar findings from HITRAN, and Barrett (2005) also made similar findings some time ago (which were predictably ignored). When these *observational* findings are added to the lack of observational evidence supporting claims of a 'climate emergency' (and as the world has warmed coming out of the Little Ice Age), using the projections of future warming for creating public policy appears extremely questionable, especially when one considers that the *models* have so far failed to agree with the observations, and are driven by extreme, unrealistic emissions scenarios.

---

2   The Arrhenius Rule states that 'if the quantity of carbonic acid increases in geometric progression, the augmentation of the temperature will increase nearly in arithmetic progression' (Arrhenius 1896, p. 70)

The Little Ice Age itself presents perhaps the biggest observational challenge to the theory that increasing concentrations of CO_2 in the atmosphere will lead to catastrophic anthropogenic global warming. This is because the recovery in temperatures began *before* there were the substantial emissions of CO_2, which accompanied the Industrial Revolution. Neither can the Roman Warm Period (when grapes were grown in Britain) nor the Medieval Warm Period (for which there is strong observational evidence, despite the best efforts of Michael Mann to render them non-existent) be explained by such a cause. (Milankovic Cycles explain the ice ages and interglacial periods, but not the smaller variations.)

There is now good evidence that the climate models run hot because they incorporate assumptions of equilibrium climate sensitivity (increases in MGT for a doubling of atmospheric CO_2) that are too high (Lewis 2023) – largely because of the assumptions of positive feedback from water vapour (by far the most powerful GHG) for which, as we noted above, there is a lack of evidence.

Neither does the observational evidence support the claims of a 'climate emergency'. As Lomborg (2020) has shown, the risk of climate-related death has declined by about 98% over the past century thanks to our global increasing economic welfare (which has also been responsible for producing the increase in GHG emissions). Moreover, Roger Pielke Jr (2023) has drawn attention to the fact that the latest IPCC report found that only on one indicator (temperature extremes) could it be concluded that observations had exceeded natural variability. Remarkably, a peer-reviewed paper confirming this IPCC conclusion (Alimonti et al. 2022) was withdrawn from publication after complaints to *The Guardian* from a number of climate scientists, including (perhaps inevitably) Michael Mann, who had notoriously dealt with the issue of natural variability by eliminating it with his 'hockey stick'.

The question that needs explanation is why climate policy has continued to be based on models that have effectively been falsified. To explain this, we can point to several political phenomena.

One causative factor that has made the climate science based on models so resilient in the face of contradictory observational evidence is

Figure 6.2 Climate-related deaths and climate-related death risk

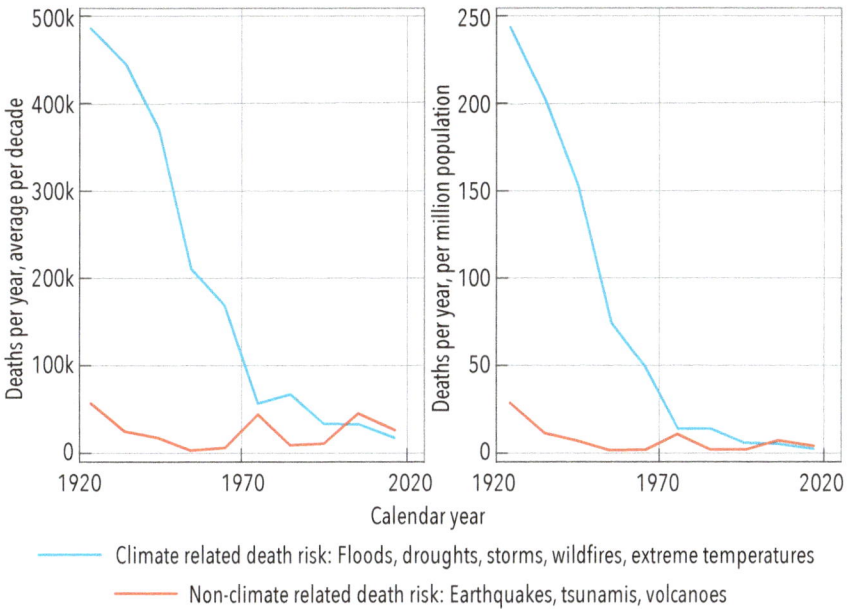

Climate related death risk: Floods, droughts, storms, wildfires, extreme temperatures

Non-climate related death risk: Earthquakes, tsunamis, volcanoes

Source: Lomborg 2020.

the existence of the IPCC. The IPCC gives an official, intergovernmental imprimatur to its conclusions, and so making possible something we might call 'international Lysenkoism' after the politicised science of genetics that had serious negative consequences for the agricultural economy of the Soviet Union. Lysenkoism is named for Soviet scientist Trofim Lysenko who held to the Lamarckian idea that characteristics acquired by an individual could be passed to subsequent generations, in contrast to the (correct) view that evolution occurs through random genetic mutations, some of which endow those inheriting them with adaptive advantage.

Mendelian genetics did not fit well with the Marxist claim that there are immutable laws of history. Stalin gave Lysenko's theory the backing of the state; consequently, dissenting scientists were marginalised and Soviet science and agriculture suffered. Any kind of 'official science' suppresses the expression of the diversity of views and evidence that is a central element of science as a process.

Cliff Ollier (2009, p. 200) showed the parallels between Lysenkoism and modern climate science: both worked through political organisations (such as the IPCC); both claimed that the science is settled (there is nothing to debate); both disregarded the accumulating evidence that the predictions are wrong; both demonised the opposition (labelling dissenters as 'deniers'); both victimised the opposition (loss of jobs or research funds); both related to a current ideology (environmentalism); both were supported by a vast propaganda machine – a huge bureaucracy where many people have careers dependent upon 'the ruling concept'.

Two features of environmentalism as an ideology assist. First, environmentalists commonly adhere to a view that Nature exists in a 'delicate balance' (Suzuki 1997), so that any variation can be both attributed to human intervention as well as portending some catastrophe if some 'tipping point' is passed. In fact, ecological science, since at least 1990, has recognised that Nature is ever changing.

Second, many environmentalists subscribe to a catastrophist world view that is commonly found in societies as a reaction to rapid change (Kellow 2020). Ellwanger (2023) points to an advantage of relying on models and their projections. Calling them projections helps mask the failure of what are really 'prophecies', and those who subscribe to policies are known to often cling to them, even in the face of evidence of their error. Leon Festinger (1957) is best known for his theory of cognitive dissonance, where people deploy all manner of defenses to fend off contrary views – such as calling their opponents 'deniers'. Interestingly, Festinger developed this insight by studying groups that had prophesied the end of the world (Festinger et al. 1956).

The opening sentences to that work could well be applied to activist climate scientists:

> A man with a conviction is a hard man to change. Tell him you disagree and he turns away. Show him facts and figures and he questions your sources. Appeal to logic and he fails to see your point. (Festinger et al. 1956, p. 3)

In addition to strong ideological factors locking in the current approach to science and policy on climate change, however, there are also powerful interests at work. The environment movement has been relentless in ascribing progress more modest than they advocate to 'fossil fuel interests' – sometimes ExxonMobil, sometimes the Koch brothers, and so on. In December 2022, climate activist Bill McKibben (2022) claimed 'front groups sponsored by the fossil-fuel industry have begun sponsoring efforts to spread misinformation about wind and solar energy'.

McKibben, who provided no evidence for this statement founded the global fossil fuel divestment movement 350.Org, and he is a beneficiary of the largesse of funders supporting climate action. In 2008, he was appointed as the Schumann Distinguished Scholar at Middlebury College, funded with a grant from the Schumann Media Center. The Schumann Foundation has a record of grants to 'pass through' foundations and groups such as the Tides Foundation, Tides Centre, Environmental Working Group, Union of Concerned Scientists, and Natural Resources Defense Council. It was instrumental in the establishment of Covering Climate Now, an international consortium of media outlets that coordinates news to emphasise the need for climate action.

Regardless of whether it was ever the case that fossil fuel interests had more resources than environment groups, it is certainly not the case now. As Robert Bryce (2023) has shown, the budgets of the groups and foundations active on climate change now dwarf those of fossil fuel and pro-nuclear business groups. The 2021 budgets of the top 25 pro-hydrocarbon, pro-nuclear groups totalled $990 million; those of the top 25 anti-hydrocarbon, anti-nuclear $4503 million.

The size of war chest is but one indicator of influence, but climate policy has destroyed the other. Business occupies a 'privileged position' in politics because it has structural power (Lindblom 1977). The economic welfare and jobs through which people derive their share of the economy, as well as the taxes government collects come from the activities of business. Governments are ultimately held responsible for maintaining this, so promises to invest or threats to disinvest give business considerable leverage. Climate policies in most liberal democracies welcome disinvestment and discourage investment in the sectors producing CO_2, so this source of influence has been rendered obsolete.

Moreover, policies produce their own politics (Lowi 1964; Pierson 1993; Kellow 2018) so there are now strong interests supporting decarbonisation policies that reinforce the normative arguments of environment groups in what Bruce Yandle (1983) termed the 'Bootlegger and Baptist' coalitions, that are stronger than either group alone. (Baptists favour prohibition on normative grounds; bootleggers require restrictions on competing businesses in order to make money.) A strong normative cloak also legitimises the funnelling of resources to the Baptists – which is the case with climate groups and economic interests using 'pass through' or 'donor advised' funds. The most egregious example is probably Sir Christopher Hohn donating £200,000 to Extinction Rebellion for its opposition to a new runway at Heathrow Airport, and then investing in Heathrow a month later.

With journals like *Nature* and *Scientific American* and news agencies like *Associated Press*, *Agence France Presse* and *Reuters* signed up to Covering Climate Now, and well-funded climate groups defending the consensus with rhetorical devices such as calling dissenting voices 'deniers' – deliberately likening them to Holocaust deniers – the groupthink is maintained. *Scientific American,* for example, reported the high temperatures of 2023, but failed to report the possible role of Hunga Tonga. The groupthink often deviates from the IPCC consensus, but the institutionalisation of that consensus in the IPCC ensures that we effectively have a situation of international 'Climate Lysenkoism'.

Just as Lysenkoism in genetics did considerable harm in the Soviet Union, Climate Lysenkoism, by exaggerating the risks of mild global warming into a catastrophic 'climate emergency' is doing considerable harm – not just economically in poor policy choices, but to children who fear that they will not have a future. There is an urgent need to open climate science up to dissent and disputation, to overcome the current 'madness of crowds'. Such a return to the proper conduct of science is likely to reveal the likelihood of modest climate change at best.

7 Precipitation: The Achilles Heel of Climate Models

Emeritus Professor David R. Legates

When you think of global warming, or in the words of the new catch-phrase, global burning, you immediately think of air temperature. Even their predecessor, global cooling, led to the belief that climate change was synonymous with a change in air temperature. Climate, however, is defined by a number of variables – some more important than others – and not simply by the thermal state of the atmosphere at any given time, location, or altitude.

Historically, two variables have stood out among the others as representing the climate of any given location: air temperature and precipitation. Their pre-eminence over other variables is due to both their direct importance on, and relationship to, issues associated with everyday life. They are also the most frequently measured variables, which in turn is facilitated by their relative ease of measurement.

Except for isolated studies, observations of air temperature and precipitation are far more prevalent than, say, measurements of humidity, winds, or solar radiation; this is because of the rather ubiquitous observations of air temperature and the higher spatial variability in precipitation. Indeed, most climate classification schemes use only air temperature and precipitation to delineate one climate type from another.

Precipitation and climate change

Given the importance of air temperature, it must be a major focus when discussing climate change trends and predictions. And, indeed, it is, but why isn't precipitation also presented? Usually, it is simply stated that a warming world is a more variable world in terms of precipitation (although it isn't), in that more floods and droughts will occur (they probably will but that is due largely to changes in land use and land cover). But maps and graphs of precipitation patterns and trends are rarely portrayed on websites and in the media, with good reason.

It is important to understand how precipitation occurs in order to answer how precipitation patterns will change in a warming world. At the macro-scale, processes that lead to precipitation require three ingredients: moisture in the atmosphere (humidity); a mechanism to cause this moisture to condense into cloud droplets; and a further mechanism to allow these cloud droplets to coalesce into ice or snow particles, or rain droplets. Having more moisture in the atmosphere only makes it more humid while condensation only leads to the formation of clouds. However, not all clouds will result in precipitation, so the processes that cause cloud particles to coalesce together to form rain droplets or snow-flakes are also required.

Without delving into cloud microphysics, only four types of precipitation-forming processes exist. All of these require the air to cool below the dew point temperature (that is, the saturation temperature), almost always by forcing the air to rise and cool adiabatically – that is, through the decrease in pressure with altitude. Thus, air is forced to rise by: cyclonic rotation (low pressure); surface heating (convection); a forced flow over topography (orographic); and interaction with air masses having different thermal and moisture characteristics (frontal) (see Figure 7.1).

How are these four precipitation-forming mechanisms likely to change under a warming world? When it is saturated, warmer air will hold more moisture than drier air – which illustrates the point that cooling the air to below the dew point temperature will lead to condensation. Thus, the atmosphere in a warmer world will potentially have a higher moisture content (specific humidity – Byrne & O'Gorman 2018). This is important to the discussion about the four precipitation-forming mechanisms, below.

Global circulation is driven by the temperature gradient between the warm Equator and the cold Poles. But if the planet is warming, the air at

Figure 7.1 Cyclonic, convective, orographic, and frontal precipitation mechanisms

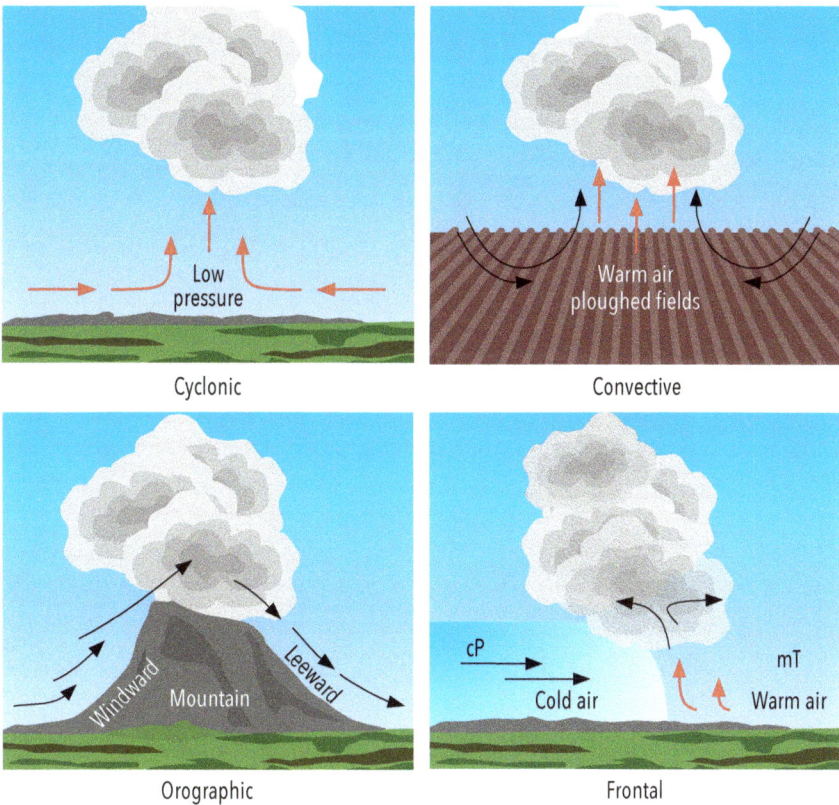

Source: Adapted from Christopherson and Birkeland 2015.

the poles will warm faster than air at the Equator due to several factors, which include:

- Colder air warms more than warmer air for the same amount of energy input (i.e., $dT \propto T^{-3}$, where T is the air temperature in Kelvins), due to the derivative of the Stefan–Boltzmann Radiation Law.
- Moist tropical air has a higher specific heat (the amount of energy required to raise the temperature of 1 gram of a substance by 1 K) than dry polar air, since water vapour has a higher specific heat than dry air.
- The change in albedo (surface reflectance) is greater in the polar regions because of the melting of highly reflective ice and snow, and the uncovering of darker soils and tundra.

- Sea ice provides a layer of insulation between the unfrozen water beneath and the potentially much colder air above. Warming reduces the sea-ice coverage, which allows energy in the relatively warmer water to warm the air above it.
- The lack of convection at the Poles keeps the warming closer to the surface. Unlike the tropics, where the atmosphere becomes unstable due to surface heating and rising air, the polar regions are not heated sufficiently to create rising air and the creation of the Polar High inhibits vertical motion. Thus, the warmer air stays near the ground.
- The evaporation of water stores energy as *latent heat* (energy in the phase transition of water from liquid to gas) which then is transported poleward by global circulation. This latent energy is released as condensation occurs in higher latitudes, thereby transporting energy polewards.

Most of these factors will lead to more warming in the winter and less during the summer. However, these factors are not limited to warming by just greenhouse gases; any cause that can be posited for a warming planet will result in an enhanced warming of polar regions.

Consider now an extreme scenario whereby warming causes the Poles and the Equator to attain the same temperature. What, then, happens to the global circulation? Since global circulation is driven by the Equator-to-Pole gradient, which is zero in this hypothetical scenario, there will be no global circulation. No Hadley cells, no trade winds, no mid-latitude westerlies, no polar easterlies, and no polar cells. Air will still move due to local spatial gradients in air temperature and pressure (for example, anabatic and katabatic winds, land and sea breezes), but the global-scale circulation will cease. Thus, in a warmed world where the Equator-to-Pole temperature gradient is reduced, global circulation will be diminished.

Cyclonic precipitation is induced by rising air motions associated with an anti-clockwise flow of air in the Northern Hemisphere and a clockwise flow in the Southern Hemisphere. Most rotational flows of air are induced by the meandering of the jet stream in the mid-latitudes, which is directly related to the Equator-to-Pole temperature gradient.

Thus, if the Equator-to-Pole temperature gradient is reduced, jet stream meanders also are reduced in their intensity, which will diminish the effect of cyclonic precipitation. By contrast, the cyclonic flow in the tropics is induced by easterly waves or 'a disturbance in the easterlies'. But as shown above, diminishing the Equator-to-Pole temperature gradient will reduce the effect of global circulation patterns, including the tropical easterlies, and the pressure difference between the low pressure at the Equator and the high pressure of the subtropics (which are 30° north or south latitudes). Consequently, cyclonic flow will be diminished both in middle and low latitudes.

Surface heating will still induce convection in a warmed world but due to a relaxed Equator-to-Pole temperature gradient, convective initiation will be diminished. Thunderstorms are less likely to develop, the frequency of severe weather (for example, large hail and tornadoes) will be diminished, and convective-induced precipitation is likely to be decreased (Hoogewind et al. 2017; Taszarek et al. 2021), particularly during the summer months. Thus, convective precipitation should decrease as well.

Regarding orographic precipitation (that is when air is forced to rise due to topography), land still provides an obstruction to air flow and the presence of anabatic or katabatic winds will enhance or diminish rising motions, respectively. However, the diminished effect of the trade winds and the mid-latitude westerlies will reduce the precipitation-causing mechanism that induces orographic precipitation by removing the persistent upslope rain or snowfall, and the concomitant rain-shadow effect on the leeward side.

Frontal precipitation results from the collision of air masses (that is, large bodies of air) with different temperature and moisture characteristics. Since warm air is less dense than cold air, and since moist air is less dense than dry air, warm moist air rises when forced by contact with cold, dry air. In the winter months, cold dry air is usually formed near the 60°north and south latitudes over land, while warm, moist air is formed over warm bodies of water near the 30° north and south latitudes. However, if the Equator-to-Pole temperature gradient is reduced, then the contrast between warm and cold air will be diminished and the lifting mechanism that creates frontal precipitation will be reduced.

If the planet warms, the Equator-to-Pole temperature gradient will decrease. While this creates a potential increase in the moisture content of the atmosphere due to the warmer air, the mechanisms required to create rainfall will be diminished. This is shown by the high humidity that exists near the Persian Gulf despite low annual precipitation totals. The ultimate question is, if the moisture content of the air increases but the magnitude and frequency of the lifting mechanisms decrease, will precipitation increase or decrease in a warming world?

Climate models and precipitation

An answer to the above question might be obtained using climate models. But as Legates (2014) noted, the simulation of precipitation within a global climate model (GCM) is quite problematic. This is because the evaporation–condensation process stores and transports energy. Thus, any error made within a GCM adversely affects the simulation of precipitation and, consequently, errors in simulating precipitation affect every other variable. For example, the moisture content of the atmosphere affects the energy released when condensation occurs and affects the formation of clouds; all of which affects how air temperature is simulated. But errors in the simulation of air temperature affect many variables, including the direction and intensity of winds, centres and magnitude of pressure patterns, and the degree of convection and storm development. Ocean circulation, too, is affected since the upper layers of the ocean are forced by wind-driven effects. Precipitation, or a lack thereof, affects the soil moisture content, which in turn impacts plant growth and the evapotranspiration rate. As a consequence, the uncertainties in modelling precipitation adversely affect every other variable in the complex interaction among the myriad of climate variables. Ferguglia et al. (2023) agree, noting that precipitation is one of the variables that is most affected by uncertainties in GCMs and these problems have been insufficiently analysed.

Given four mechanisms can lead to precipitation, it is imperative that the reasons for changes in precipitation are correctly determined. Whether precipitation increases due to an increase in cyclonic precipitation or because of an enhancement of frontal precipitation events, for example, is important. Moreover, it also is important to know *why* these patterns will potentially change. Is it due to a change in atmospheric circulation, more (or less) moisture in the atmosphere, variations in the vertical structure of atmospheric temperature, or something else? And correctly identifying the mechanism is key to making a useful prediction for our future climate. A climate model can be tuned so that the spatial and temporal distribution of air temperature appears reasonable (Golaz, Horowitz & Levy 2013; Suzuki, Golaz & Stephens 2013; Dommenget & Rezny 2017; Hourdin et al. 2017; Faghih et al. 2022), but what effect does tuning have on our simulation of precipitation? Will we get a reasonable answer that is forced by inappropriate physics? How will we know whether the results are reasonable and valid?

One of the many problems associated with precipitation simulation within a GCM is that only one of the four mechanisms – surface convection – produces virtually all of the precipitation in the model. Precipitation within a GCM occurs relatively frequently over large regions and resembles the popping of popcorn kernels when heated (Zolina 2014), producing rainfall that occurs too often but at an intensity that is too low (Stephens et al. 2010; Martinez-Villalobos et al. 2022). When averaged, the general global pattern of precipitation emerges (for example, wet rainforests, dry deserts and polar regions), although too much precipitation is produced over the tropical oceans and too little in middle latitudes (Stephens et al. 2010). It may seem that the model correctly represents precipitation patterns, but in fact the model may yield a reasonable pattern for the wrong reasons. Differences between the models and observations are much greater than the uncertainty in measuring precipitation (Stephens et al. 2010). Even if total precipitation appears reasonable, it is usually for the wrong reasons.

Another fundamental problem with the simulation of precipitation within GCMs is that they exhibit considerable intra-model variability, but lack inter-annual variability. Simply put, the models vary considerably among themselves but each year within the model looks very much like every other year. Several researchers have demonstrated that regional estimates of precipitation vary widely from model to model (Dai 2006; Stephens et al. 2010; IPCC 2013; Kim et al. 2021; Thomasson et al. 2021;

Vicente-Serrano et al. 2022). Christy (2012) discusses the variability in the southeastern and midwestern United States where the models differ by as much as a factor of two with significant underestimation of precipitation in the southeast and overestimation of the midwest and the southwest (see also Langford et al. 2014). Even the daily cycle of precipitation, particularly the terrestrial nighttime peak in rainfall, is not well simulated by state-of-the-art GCMs (Lee & Wang 2021; Zhou et al. 2023).

In addition to the inability of GCMs to simulate the processes that cause precipitation and lead to its variability, models are unable to simulate the intra-annual variability in precipitation. Soden (2000) writes that 'not only do the GCMs differ with respect to the observations, but the models also lack coherence among themselves ... even the extreme models exhibit markedly less precipitation variability than observed' and concludes, 'if the GCMs are in error, this deficiency would

Figure 7.2 An example of the spread among different Climate Model Intercomparison Project (CMIP) climate models

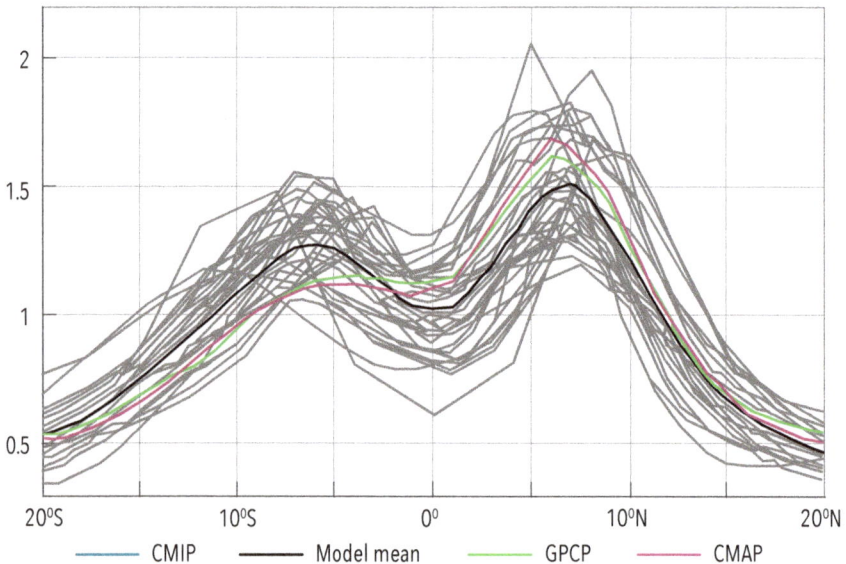

Source: Kim et al. 2021. The graph shows the annual- and zonal-mean tropical precipitation normalised by tropical-mean for different CMIP climate models (grey lines); the multi-model mean (black line), and observations from the Global Precipitation Climatology Project (GPCP, green line) and the CPC (Climate Prediction Center) Merged Analysis of Precipitation (CMAP, pink line) datasets.

presumably reflect a more fundamental flaw common to all models'. Other authors have noted this marked underestimation of the observed inter-annual variability (Wang et al. 2011; Song & Zhou 2014; Guan et al. 2022; Martinez-Villalobos et al. 2022; Xu et al. 2022).

GCMs also are flawed in their representation of topography, particularly in spectrally-based GCMs (that is models where fields are represented as a series of spherical harmonics rather than specific grid boxes). Lindberg & Broccoli (1996) and Biasutti et al. (2003) demonstrate that at traditionally used resolutions, the elevation of mountain ranges is underestimated and their spatial extent is overestimated; they resemble large hills more than jagged mountains. Since spectral models represent topography using smoothly varying functions (spherical harmonics), wave trains are produced that create false mountain ranges even over the relatively flat oceans. Moreover, model simulations of precipitation vary significantly as a result of the horizontal resolution used in the model, owing, in part, to the variation in topography (Done et al. 2005; Gutowski et al. 2020; Kim et al. 2022; Light et al. 2022; Wang et al. 2023). Including the uncertainty in precipitation is critical in model confidence estimates for policy-making (Uhe et al. 2021).

Figure 7.3 Topography as represented by the National Centers for Environmental Prediction (NCEP) GCM (left) and a five-minute digital elevation model (right)

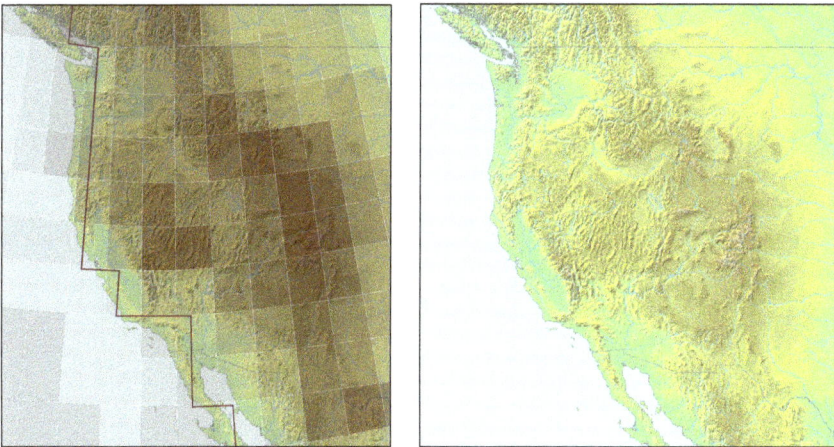

Source: Adapted from Hostetler et al. 2003.

Conclusion

Precipitation is an important but overlooked variable in climate models. The problem is that precipitation is not well represented temporally – in terms of the daily cycle as well as the inter- and intra-annual variability – or spatially. Many reasons for this exist; but since precipitation varies considerably on a scale much smaller than the smallest resolvable grid scale within GCMs and the models represent only one precipitation forming mechanism (convection), it is unlikely that this problem will be remedied soon. Moreover, any error in the model simulation adversely affects the simulation of precipitation and, in turn, errors in the simulation of precipitation affects virtually every other climate variable. Consequently, the simulation of precipitation is the Achilles Heel of climate models and is possibly the main reason presentations of the ability of the model to simulate precipitation are usually suppressed.

8 Characteristics of Recent Climate Change
William Kininmonth

Introduction

Earth's global average near-surface temperature, in the form of an anomaly from the long-term mean, is a widely used index for measuring climate change. In the following discussion it will be demonstrated that this index is a poor and even misleading descriptor. The global temperature anomaly masks regional and seasonal variations that are important for understanding the processes driving climate change.

The use of satellites for monitoring temperature has provided global coverage over land and sea, at the surface and through the atmosphere. The systematic collection and archival of analyses based on observations made over the recent four decades has provided a powerful database for analysing the changing climate. The regional and seasonal characteristics strongly suggest that natural processes have been a primary cause for recent warming.

Characteristics of recent temperature change

The following analyses use data collected from 1979 through 2022, processed and archived as NCEP/NCAR (R1) by the US National Center for Environmental Prediction (Kalnay et al. 1996). The data are readily available online from the NOAA Physical Sciences Laboratory (NOAA n.d.).

Table 8.1 sets out the zonal trends (°C/century) for the period 1979 through to 2022 of atmospheric temperature at various pressure

Table 8.1 Temperature trend (°C/century) by latitude band and altitude

LATITUDE	ALTITUDE			
	2 metres	850 mb	700 mb	500 mb
60°N–90°N	6.8	4.9	3.5	2.7
30°N–60°N	2.5	3.5	3.0	2.8
10°N–30°N	1.6	2.1	2.4	1.5
10°S–10°N	1.1	1.5	2.1	1.0
30°S–10°S	0.5	1.7	1.4	1.4
60°S–30°S	0.2	2.2	1.8	2.0
90°S–60°S	3.1	3.3	2.4	2.0
GLOBAL	1.7	2.5	2.3	1.8

Source: Data from NCEP/NCAR (R1) for the period 1979–2022.

altitudes.[1] The global near-surface (2 metre) temperature trend is 1.7 °C/century, but this value masks significant zonal differences. The warming for the equatorial band (latitude 10°S to latitude 10°N) is only 1.1 °C/century, whereas the trend for the Arctic (north of latitude 60°N) is 6.8 °C/century and that of the Antarctic (south of latitude 60°S) is 3.1 °C/century. The relatively strong warming over the Poles persists through the lower troposphere up to 500 millibars (mb).

Further differences appear when the regional temperature trends are analysed by seasons. Table 8.2 separates out the summer and winter trends for the polar regions.

Over the Arctic, the 8.1 °C/century near-surface temperature trend for the months of darkness (October to March) is much greater than the 4.2 °C/century trend for the months of sunlight (April to September). Although the warming is strong during all months from the autumn to spring equinox, the peak warming of 12.6 °C/century is in October. In contrast, the warming trend is only 0.7 °C/century during June as solar radiation intensity is approaching the summer peak.

---

1   The altitudes of the pressure levels are approximately as follows: 850 mb – 1.5 km; 700 mb – 3.0 km; 500 mb – 5.5 km.

Table 8.2 Near surface (2 metre) temperature trends (°C/century) by season over the Arctic (latitude 60°N–90°N) and Antarctic (latitude 90°S–60°S)

	SEASON	
	WINTER	SUMMER
ARCTIC	October – March 8.1	April – September 4.2
ANTARCTIC	April – September 5.2	October – March 1.2

Source: Data from NCEP/NCAR (R1) for the period 1979–2022.

A pattern of seasonal differences is also observed in the Antarctic data, with a near-surface warming trend of 5.2 °C/century during the darker months (April to September) but a much smaller trend of 1.2 °C for the sunlight months (October to March).

In contrast to the polar regions, there are no seasonal differences for equatorial temperature trends.

The regional trends are significantly different from the global average. These differences are masked by using the metric of global average temperature anomaly as an index of climate change. Any explanation for recent warming must acknowledge and be able to accommodate these regional and altitudinal differences.

Ocean temperature forcing

Data from the satellite era demonstrate that atmospheric temperatures are responding to changes in equatorial ocean surface temperature. Figure 8.1 is a chart of monthly ocean surface and 700 mb temperature anomalies for the equatorial band, latitudes 10°S to 10°N. The use of the anomaly is to remove the annual cycle from the visual presentation but does not alter the derived statistics. The equatorial band covers the equatorial trough where atmospheric convection is most active. Here, the range of year-to-year variation of ocean surface temperature is nearly 2 °C, with positive departures associated with El Niño events and negative departures associated with La Niña events.

The atmospheric temperature closely follows the ocean surface. The detrended correlation between the two series is 0.70. The correlation

Figure 8.1 Anomalies of temperature from the seasonal mean at the ocean surface (SST) and 700 mb altitude for the equatorial band (latitude 10°S to 10°N)

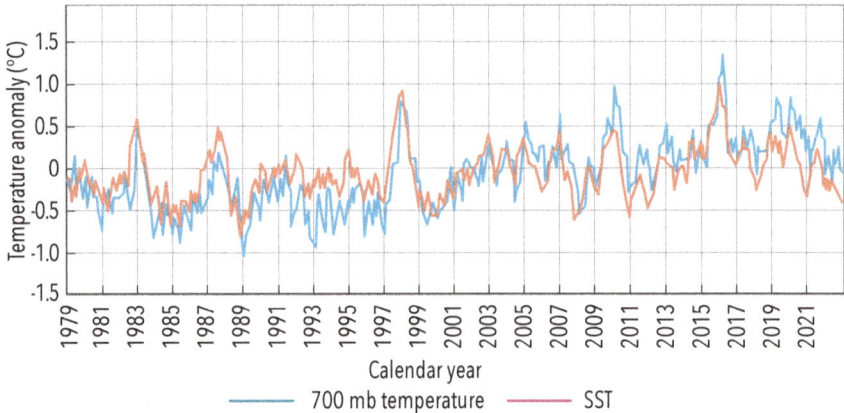

Source: Data from Kalnay et al. (1996) NCEP/NCAR (R1) for the period 1979–2022. The long-term trend of the ocean surface temperature is 0.7 °C/century and at 700 mb is 2.1 °C/century.

increases to 0.76 when the atmospheric temperature is lagged by one month, and to 0.79 when it is two months lag. That is, changes in atmospheric temperature follow changes of ocean surface temperature.

Equatorial ocean temperature regulates the atmospheric temperature because over the tropics solar radiation passes through the atmosphere and is absorbed in the surface layer of the ocean. Solar radiation warms the ocean but not the atmosphere. Energy in the tropical ocean surface layer flows to the atmosphere as heat and latent heat, where latent heat is contained in evaporating water vapour.

Over the tropics, the heat and latent heat accumulating in the lower troposphere are advected by the Trade Winds to the equatorial trough (Trenberth & Stepaniak 2003). Within deep convection clouds, particularly over the equatorial trough region, air rises buoyantly through the atmosphere, some reaching the tropopause at about 12 km altitude. The rising air in convection clouds cools at a specific rate known as the moist adiabatic lapse rate. Some of the rising air mixes with the surrounding air. Consequently, the temperature of air surrounding the convection clouds changes with the ocean surface temperature.

Underlying the interannual variations noted in Figure 8.1 there are also long-term trends. The equatorial ocean surface trend is 0.7 °C/century whereas at 700 mb altitude the trend is 2.1 °C/century. The reason for the difference is that the moist adiabatic lapse rate is not constant. As surface temperature increases, so too the lapse rate trajectory 'spreads' with altitude. The rate of surface temperature increase is amplified at altitude. Consequently, as the equatorial ocean surface temperature has risen with time then the atmospheric temperature has warmed at a slightly greater rate.

Energy flow and climate change

To understand the processes behind the temperature trends and regional differences identified in Tables 8.1 and 8.2, it is necessary to follow the flow of energy from the tropics to higher latitudes. There are discrete energy exchange and transport components to the pathway, each contributing to local changes in different ways.

The role of greenhouse gases in the atmosphere (water vapour and CO_2, plus other minor gases) is often presented as absorbing radiation emitted from the surface to keep the atmosphere (and Earth) warm. This is false. The greenhouse gases do absorb radiation emitted from the surface, but they also emit radiation, both to space and back to the surface. The emissions to space and back to the surface exceed the absorption; the greenhouse gases tend to cool the atmosphere (Kiehl & Trenberth 1997).

The radiation loss does not actually cool the atmosphere because the energy lost is replaced by the flow of heat and latent heat from the surface. The rates of flow are regulated by surface temperature: the warmer the ocean surface temperature, the faster the flow of energy.

The tropical surface temperature has warmed over recent decades and the rate of flow of energy from the ocean to the atmosphere has increased. However, it is latent heat that dominates the tropical energy flow, and this energy is not immediately available as heat. Within convection clouds, heat and latent heat of rising air are transformed to potential energy in the upper troposphere (Riehl & Malkus 1958).

As the winds of the upper troposphere transport air from the tropics to middle and high latitudes, the potential energy of the air is also transported polewards. Away from the equatorial trough, air subsides to compensate for the convection. As air loses altitude the potential energy is transformed to heat. It is this heat that is available to offset the loss of energy from radiation. If the overturning rate of the circulation were to increase then, away from the equatorial trough, more heat would be generated to warm the atmosphere.

The equator to pole temperature difference is strongest during the winter months. Consequently, the poleward transport of potential energy, and its realisation as heat in the troposphere, is also strongest in the winter hemisphere (Trenberth & Stepaniak 2003).

The construct of energy flow outlined above provides a clear link between the slowly warming tropical ocean and the observed wintertime warming over polar latitudes. Heat and latent heat flow from the warm tropical oceans to the air near the surface. The heat and latent heat are transported to the equatorial trough in the trade winds where the near surface air is caught in the buoyantly rising air of convection clouds. Within the clouds, heat and latent heat are transformed to potential energy in the upper troposphere. The winds of the upper troposphere transport air poleward, particularly during the winter. In the subsiding air, potential energy is realised as heat and offsets radiation energy loss. Any warming of the tropical ocean will increase the flow of energy to the atmosphere, further increasing the wintertime release of heat to the atmosphere over middle and high latitudes.

Ocean warming

A fundamental question remains: why has the tropical ocean warmed in recent times?

There is no evidence of a significant increase in solar intensity over recent decades. Recognising the solar cycles, the input of solar energy to the oceans has effectively been steady over the period.

The increasing concentration of CO_2 in the atmosphere has been widely promoted as a basis for recent global warming. However, as noted, radiation emitted by greenhouse gases tends to cool the atmosphere.

Where the greenhouse gases do interact with Earth's energy flow and can modify surface temperature is through the emission of long-wave radiation back to the surface. Absorption of this long-wave radiation has reduced net long-wave radiation loss from the surface and raised the surface temperatures.

Table 8.3 indicates that over recent times the increase in CO_2 concentration has had negligible impact on long-wave radiation absorbed by the tropical surface. Table 8.3 is constructed from the MODTRAN radiation transfer model[2] using clear sky tropical temperature and moisture profiles, but with changing concentrations of CO_2. With the standard water vapour profile and no CO_2, the long-wave radiation reaching the surface from the atmosphere is 361.40 W/m². At the Last Glacial Maximum (LGM), 20,000 years ago, the CO_2 concentration was about 200 parts per million (ppm) and contributed an additional 6.61 W/m² to the absorbed surface long-wave radiation.

From the Glacial Maximum to the Holocene, 10,000 years ago, the CO_2 concentration increased by about 100 ppm and the long-wave radiation reaching the surface increased only by 0.63 W/m². The CO_2 concentration increased by a further 100 ppm during the recent period of global industrialisation, but the further increase in long-wave radiation reaching the tropical surface was only 0.62 W/m². The absorption of

Table 8.3 Incremental increases in long-wave radiation absorbed at the tropical surface as carbon dioxide concentration increases

	Long-wave radiation absorbed at the surface from water vapour and CO_2				
Carbon Dioxide (pm)	0	200	300	400	600
Surface Radiation (W/m²)	361.40	368.01	368.64	369.26	370.25
Incremental Increase		6.61	0.63	0.62	0.95

Source: Calculations using the MODTRAN medium resolution radiation transfer model for the tropical atmosphere, under clear sky conditions and with standard tropical temperature and water vapour profiles.

2   MODTRAN is a medium resolution radiation transfer model available through the University of Chicago at MODTRAN Infrared Light in the Atmosphere (uchicago.edu)

long-wave radiation at the tropical surface is dominated by the abundant water vapour. Changing CO_2 concentration has little effect on tropical ocean surface temperatures.

It is salutary to reflect on how Earth's climate has changed during these incremental increases in CO_2 concentration. From the LGM to the Holocene there was dramatic climate change, including melting of polar ice sheets and mountain glaciers sufficient to raise sea level by about 130 metres. During the recent period of industrialisation, the incremental CO_2 increase has been similar, but without systematic observations it would be difficult to identify any change in sea level.

A more plausible explanation for recent warming of the tropical ocean surface temperature relates to changing ocean currents. The dramatic changes in regional climate during El Niño and La Niña events is well documented and understood. Regional climate impacts emanate from changes linked to ocean currents of the equatorial Pacific Ocean. Changing the large-scale ocean circulations cannot be discounted from consideration.

Over the tropics, Earth absorbs more solar radiation than the long-wave emission to space. In contrast, over middle and high latitudes the emission to space is greater than the solar absorption (Trenberth & Caron 2001). A global near steady state balance between absorbed solar and emitted long-wave radiation is only achieved through poleward transport of energy by winds and ocean currents.

The absorption of solar radiation by the surface layer of the tropical ocean and the flow of heat and latent heat to the atmosphere has been discussed. However, part of the absorbed solar radiation is transported by the ocean currents to higher latitudes. From conservation of energy principles, a slower rate of poleward transport by ocean currents results in warming of the tropical ocean and an increase in the rate of flow of heat and latent heat to the tropical atmosphere.

A recent slowing of the North Atlantic Gulf Stream (Caeser et al. 2018), a major contributor to ocean heat transport, would explain the observed warming of tropical ocean surface temperatures and the associated warming of the atmosphere.

Conclusion

Analysis of data collected over the recent four decades provides a clear pattern of warming, which is masked in the metric of global average temperature widely used as an index to describe climate change. The energy source of the observed atmospheric warming is an increased rate of flow of heat and latent heat from the tropical oceans to the atmosphere as the oceans have slowly warmed. The wintertime polar amplification of warming has occurred because of the increased rate of input of latent heat from the tropical oceans and enhancement of the natural seasonal cycle of poleward energy transport.

The strong seasonality of warming over middle and high latitudes means that the springtime thaw is occurring earlier and the autumn freeze later. Summer ice melt is occurring over a longer duration, causing land ice to melt and mountain glaciers to retreat.

The earlier springtime thaw and later autumn freeze have also extended the growing seasons over middle and high latitudes. Analysis of satellite data suggests there has been a greening of the planet over recent decades, together with steadily increasing food production. Although food production has benefitted from technological changes, improved plant strains, and increasing atmospheric CO_2 to aid photosynthesis, the longer growing season has been the major contributing factor to increased vigour of natural ecosystems.

The reason for the slow warming of the tropical oceans remains an open question. However, the warming is unlikely to have been due to changing solar intensity or the increase in CO_2 concentration in the atmosphere. The more likely explanation for the observed tropical ocean warming is a decrease in poleward heat transport by the ocean currents, including a slowing of the North Atlantic Gulf Stream. A decrease in the rate of ocean heat transport and a commensurate increase in tropical ocean surface temperature are consistent with the pattern of latitudinal and seasonal rates of warming of the atmosphere observed over recent decades.

9 What Causes Increasing Greenhouse Gases?

Hermann Harde

Introduction

The Intergovernmental Panel on Climate Change (IPCC) classifies the human influence on our climate as extremely likely to be the main reason for global warming over the last few decades (IPCC 2021). In particular, anthropogenic emissions of carbon compounds, with carbon dioxide (CO_2) as the main culprit and methane (CH_4) as a distant second, are made responsible for the observed temperature changes, while any natural forcings are almost completely excluded.

This chapter addresses the question, how much human or native emissions can be made responsible for the observed increase of greenhouse gases (GHG), considering, in particular, the concentration of CO_2 in the atmosphere. It is based primarily on current studies (Harde & Salby 2021; Salby & Harde 2021a; Salby & Harde 2021b; Salby & Harde 2022), which arose from a close cooperation with the late Murry Salby, whose work on the carbon cycle – already more than ten years ago – raised significant doubts about the IPCC's explanation, particularly the assertion of an exclusively man-made increase of CO_2.[1]

---

1   Murry Salby's criticism of the IPCC's view led to a dissention with his faculty members at Macquarie University, Sydney, and ultimately to his early dismissal. The great personal attacks and existential impairments Professor Salby had to bear over recent years are certainly one reason of his early death.

In times that are more determined by pseudo-science than serious climate science we have lost an important voice standing up for the true values in science and research. We pay the greatest tribute to the deceased for his extensive scientific work, which has contributed to a much deeper understanding of climate processes.

Discussed in the following sections are: the results of an in-depth analysis on the record of atmospheric $^{14}CO_2$, an isotopic tracer of CO_2, to better understand how CO_2 is removed from the atmosphere; an examination of the close correlation of tropical warming and net CO_2 emissions as an important natural influence that likewise figures in the control of atmospheric CO_2; and a discussion about the physical mechanisms through which observed warming can produce the observed evolution of CO_2.

The relationship of carbon 14 to the removal of CO_2

The observed CO_2 evolution, inclusive of its annual cycle, has recently been reproduced in numerical simulations (Harde & Salby 2021; Harde 2017; Harde 2019; Berry 2021; Harde 2023). They show, how the abundance of CO_2 in the atmosphere is controlled by a competition between two opposing influences – the feed of CO_2 through emission, and its removal through absorption – both at the Earth's surface. This competition governs time-mean CO_2, where absorption figures centrally. It determines if and how fast CO_2 grows, as well as the magnitude of its perturbation, for example, by anthropogenic emissions. Yet, actual observations of CO_2 absorption are scarce. However, the impact of global absorption on atmospheric CO_2 is represented in carbon 14 (^{14}C), an isotope of atmospheric carbon that has been observed in the troposphere since the 1950s (CDIAC 2017).

Carbon 14 has a radioactive decay time of 8267 years (e-folding time). On time scales of relevance, the operation of $^{14}CO_2$ is virtually identical to that of the preponderance of CO_2 isotopologues with 98.9% of $^{12}CO_2$ and 1.1% of $^{13}CO_2$, those comprised of the stable isotopes ^{12}C and ^{13}C. Dynamical, chemical, and thermodynamic processes acting on those three isotopes of CO_2 (including those in the biosphere) are, for practical considerations, indistinguishable.

This feature makes ^{14}C a unique tracer of atmospheric CO_2 and provides an unrivalled means, through which to understand key mechanisms controlling the evolution of atmospheric CO_2. Once CO_2 is introduced into the atmosphere, whatever influence is experienced by one isotopologue is experienced by the others. Owing to this property and its artificial enrichment by nuclear testing, ^{14}C is central to estimates of CO_2 absorption, which vary widely. Absorption, in turn, is essential in understanding changes of atmospheric CO_2.

During the 1950s and early 1960s, atmospheric testing of nuclear devices sharply enriched ^{14}C in the stratosphere. Through atmospheric circulation, ^{14}C-enriched air in the stratosphere was subsequently transferred into the troposphere. By 1963, when the Nuclear Test Ban Treaty (NTBT) was implemented, tropospheric ^{14}C had increased by nearly 100%. The NTBT virtually eliminated the anomalous nuclear source of ^{14}C, leaving its perturbation of ^{14}C to decline through absorption and overall ^{14}C to return to its nearly unperturbed equilibrium abundance.

Figure 9.1 displays a measurement of anomalous $(\Delta^{14}C)_C$ at Vermunt, Austria (Levin et al. 1994) as fractional departure from a reference abundance and re-normalised to the atmospheric CO_2 mixing ratio at 1959 (yellow line, see Harde & Salby 2021).

An in-depth analysis of the ^{14}C record reveals that in addition to long-term behaviour it is also important to consider short-term changes

Figure 9.1 Observed $\Delta^{14}C$ at Vermunt, Austria

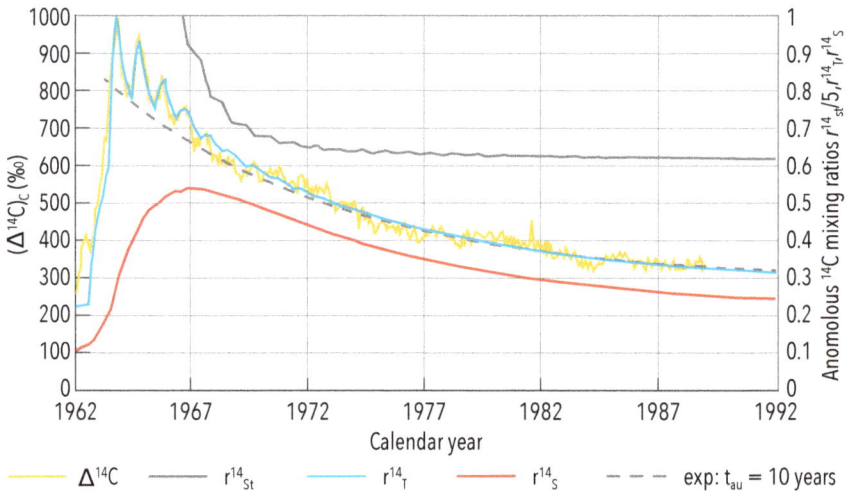

Source: Levin et al. 1994. Observed $\Delta^{14}C$ corrected for an increasing CO_2 mixing ratio (yellow), together with the relative mixing ratios r_{St}^{14}, r_T^{14} and r_S^{14} of anomalous ^{14}C in the stratosphere (grey), troposphere (blue) and surface layer (red). An exponential decay with an e-folding time of ten years is shown as a grey dashed line.

that have been largely ignored. Those changes exhibit the underlying mechanisms responsible for the observed decline of atmospheric $^{14}CO_2$ and, thereby, for removal of overall CO_2. They represent absorption that is considerably faster than appears in the average decline of $^{14}CO_2$, initially and also later in its long-term decline.

Such analysis considers the response to impulsive emission in a three-volume system (Salby & Harde 2021a, Eq. 10): the stratosphere, the troposphere, and the surface layer. The latter is characteristic of the ocean mixed layer and the uppermost layer of land, both surface layers that exchange carbon with the troposphere. With a carbon mass in the troposphere, nine to ten times that of the stratosphere and approximately one quarter of the surface layer, any carbon mass exchange between the volumes causes quite different changes in the anomalous $^{14}CO_2$-concentrations or mixing ratios r_{St}^{14}, r_T^{14} and r_S^{14}. So, to observe a noticeable change in r_T^{14} this presupposes a much larger initial value and change of r_{St}^{14}.

When solving the coupled balance equations of the three-volume system for the anomalous mixing ratios, r_{St}^{14} in the stratosphere (grey line, divided by five to fit to the scale), r_T^{14} in the troposphere (blue line) and r_S^{14} on the Earth's surface (red line), this is relative to the maximum tropospheric concentration, the observed evolution of $^{14}CO_2$ in the troposphere can be reproduced well, for both long and short time scales.

The enriched ^{14}C concentration in the stratosphere is mainly transferred to the troposphere by annual short pulses due to the Brewer–Dobson circulation, especially during boreal winters, whereas an exchange driven by the contrast $r_{St}^{14} - r_T^{14}$ is of the order of 10 to 20 years. In the troposphere, CO_2 is homogenised hemispherically in only a couple of weeks. The declines following the annual re-enrichments, track exponential declines with a direct absorption time τ of $0.5 - 2.0$ years. Absorption during those intervals is an order of magnitude faster than is reflected in the average decay (grey dashed line with a decay time of ten years) as well as in the long-term decline that is controlled by the removal through sequestration and/or dilution beneath the surface layer. It prevails later with an e-folding time of $\tau_S \approx 7.7$ years (red line).

Despite the fast exchange between the troposphere and surface layer with a direct absorption time of $\tau \approx$ eight months and an exchange rate – $(r_T^{14} - r_S^{14})/\tau$, is the effective absorption of atmospheric CO_2 determined by the direct absorption rate r_T^{14}/τ and the re-emission r_S^{14}/τ from the surface to the troposphere. Expressing this re-emission as fraction β of the direct absorption with $r_S^{14}/\tau = \beta \cdot r_T^{14}/\tau$, the exchange rate converts to $-r_T^{14}/\tau_{eff}$ with an effective absorption time $\tau_{eff} = \tau/(1 - \beta)$. Depending on the degree of re-emission of previously absorbed CO_2 back from the Earth's surface to the atmosphere, τ_{eff} can be as short as the direct absorption time or approach to the removal time τ_S of the surface layer.

So, the average decline of $^{14}CO_2$ is slowed initially by periodic re-enrichment from the stratosphere, which offsets direct absorption at the surface. Finally, however, its decline is slowed by re-emission of absorbed $^{14}CO_2$ from the surface, which likewise offsets direct absorption.

Applying the same considerations to anthropogenic emissions of CO_2 recovers effective absorption that can even be faster than the mean decline of $^{14}CO_2$. The difference follows from a magnified disequilibrium between the atmosphere and the Earth's surface. While $^{14}CO_2$ was perturbed impulsively by nuclear testing, the absorption of anthropogenic CO_2 is perturbed continuously by fossil fuel emissions in the troposphere. Solving the same three-volume-system, now for the anomalous mixing ratios r_{St}^a, r_T^a and r_S^a, caused by the anthropogenic emissions over the Mauna Loa Era (since 1958) (CDIAC 2017) at otherwise same conditions, these continuous and slightly increasing emissions also increase the disequilibrium between the atmosphere and surface layer (see Figure 9.2). While the stratospheric concentration (orange) and tropospheric concentration (blue) are close to equilibrium, continuous emissions cause a further increasing contrast to the surface layer (red), which inhibits a larger offset of direct absorption by re-emissions and results in faster effective absorption of anthropogenic CO_2.

Figure 9.2 also shows that up to the end of 2023, anthropogenic emissions cannot have contributed more than 13 parts per million (ppm) to the atmospheric concentration, which is just 9% of the CO_2 increase since the beginning of the Industrial Era; further constant emissions over

Figure 9.2 Solution of a three-volume system for the anomalous anthropogenic mixing ratios

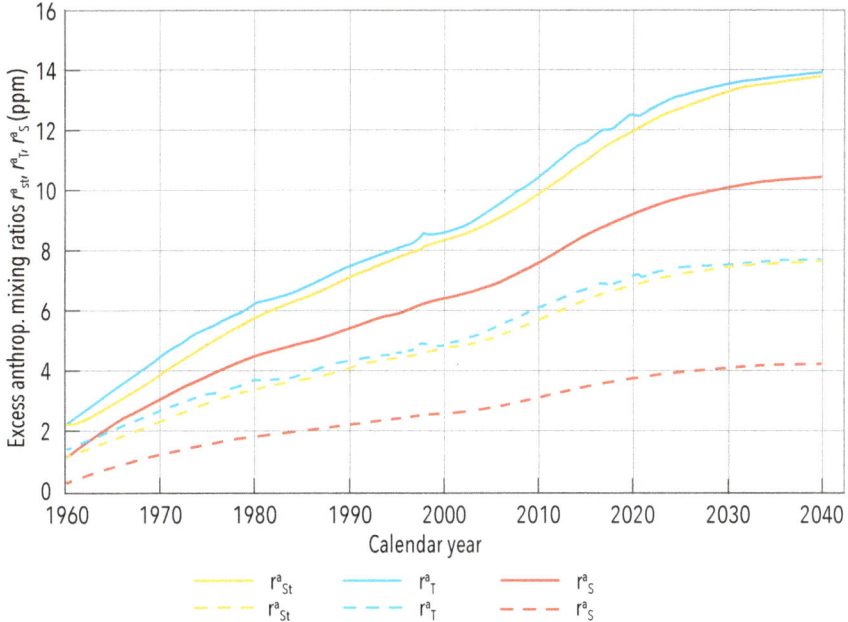

Source: CDIAC 2017. The three-volume system for the anomalous anthropogenic mixing ratios r_S, r_T and r_S are, in the stratosphere (orange), troposphere (blue) and surface layer (red) with a tropospheric to surface carbon mass ratio $m_T/m_S = 0.25$ (solid lines) and $m_T/m_S = 0.1$ (dashed lines).

successive years can only increase the concentration by an additional 1 ppm.

When assuming increased entrainment from the mixed oceanic layer into the deep ocean, or only a larger mixed layer, in the same way as the carbon mass of the extraneous reservoirs are increasing, the excess mixing ratio r_S^a of the surface layer is decreasing and, likewise, the anthropogenic contribution r_T^a in the troposphere is also decreasing. For a carbon mass ratio of the troposphere to surface layer with $m_T/m_S = 0.1$ instead of 0.25, representative for a mixed layer depth of about 100 m, for example, the respective concentrations are displayed as dashed graphs. Then the anthropogenic contribution is reducing to $r_T^a = 7.3$ ppm, which is about 5% of the observed CO_2 increase since the Industrial Era.

From the IPCC's own estimates of extraneous carbon reservoirs, the anthropogenic contribution to increased CO_2 has been shown to be no more than 15% to 35% (Harde & Salby 2021; Harde 2017; Harde 2019; Berry 2021; Harde 2023). The present analysis does not rely upon such estimates and shows that the anthropogenic perturbation must even be smaller. Altogether, the observed behavior of $^{14}CO_2$ provides an upper bound on the anthropogenic perturbation of atmospheric CO_2, which contributes not more than 10% to the observed increase.

Influence of tropical warming

The surface processes that regulate emission and absorption of CO_2 depend intrinsically upon temperature (Salby & Harde 2021b). Many, like soil respiration, even increase exponentially with temperature, typical of Arrhenius temperature dependence that operates in chemical reactions and underpins surface processes.

Global temperature today is ~0.7 °C warmer than it was half a century ago. The most reliable record of global temperature is the satellite record from the Microwave Sounding Unit (MSU) suite of instruments (Spencer et al. 2017), which retrieves temperature with homogeneous and near-global sampling of the Earth. It indicates that over much of the Earth, surface temperature underwent no systematic (ubiquitous) heating during the last four decades that was observed by the MSU. Perceptible heating was only introduced during just two brief intervals, both for not more than about two years in length: one preceding the El Niño in 1997, the other preceding the El Niño in 2016. It would be virtually impossible for such heating twenty years apart to be caused by continuously emitted anthropogenic GHGs, which are mainly released in the northern mid-latitudes.

The exception, however, is surface temperature in the tropics, where temperature systematically increased during the four decades that have been observed by the MSU. The sustained increase is also mirrored in the independent record of anomalous sea surface temperature (SST) from the Hadley Centre (Kennedy et al. 2019).

Owing to the dependence on temperature of the physical and chemical processes that regulate CO_2 emissions, CO_2 must have experienced a parallel influence. This can directly be scrutinised by investigating the interdependence of the observed temperature records and the *net* CO_2 emissions, known as E_{net}, and the component of the emissions that actually changes CO_2 and represents the difference of the total emission rate E and the absorption rate r_A/τ_{eff} with $E_{net} = E - r_A/\tau_{eff}$. It is derived from the measured CO_2 concentration r_A in the atmosphere as the instantaneous rate of change dr_A/dt and determined by the conservation law $dr_A/dt = E - r_A/\tau_{eff}$ for atmospheric CO_2.

As a widely accepted reference, we rely on the measurements of CO_2 at Mauna Loa, Hawaii (CDIAC 2017), which are largely free of local distortions, at least until November 2022. They approximate the global abundance of CO_2, which, on time scales longer than a month, is well mixed across the free atmosphere.

Plotted in Figure 9.3 is E_{net}, low-pass filtered to periods longer than a year (blue) to remove the seasonal cycles, for which mean net emission associated with this component can be shown to vanish.

Superimposed is the component of net emissions that operate coherently with temperature T_{SL} over tropical land, observed by MSU and extrapolated backwards before the satellite era through the record

Figure 9.3 Net CO_2 emission observed (blue), low-pass filtered to periods longer than a year. Superimposed are the expected emissions induced by the temperature over tropical ocean and land (red).

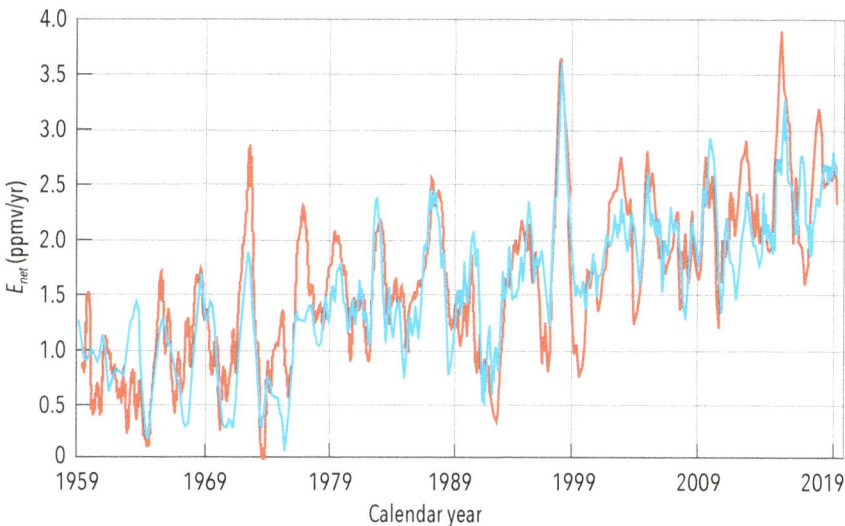

of SST (red). It is evaluated by projecting E_{net} onto the record of T_{SL} from SST and MSU. The derived record $E_{net}(T_{SL})$ closely tracks observed net emission, *Enet*. With a correlation of 0.81, their interdependence is highly significant.

Fluctuations of emissions, operating on time scales of only a couple of years, introduce anomalous CO_2 that falls within the short-time regime, that is, before re-emission can intensify and offset direct absorption. Such perturbations will, therefore, experience effective absorption that is fast – comparable to direct absorption. On the other hand, emissions that vary slowly or are invariant, will introduce anomalous CO_2 that falls within the long-time regime, when re-emission has intensified and offsets direct absorption. Such perturbations will therefore experience effective absorption that is slow, operating at only a fraction of the pace of direct absorption.

Thermally induced emissions, especially from tropical land surfaces, are found to represent much of the observed evolution of net CO_2 emissions. It accounts for the sporadic intensifications of net emissions that operate on interannual time scales, notably during the episodes of El Niños, and equally well for the long-term intensification during the last half century. Jointly, these unsteady components of net emissions determine the thermally induced component of anomalous CO_2 and closely track the observed evolution of CO_2 (see the next section).

The veracity with which the thermally induced component reproduces the observed evolution of CO_2 has two important implications:

- Tropical land temperature is a robust predictor of atmospheric CO_2. By contrast, other contributions to net emissions that operate incoherently with the temperature afford virtually no predictive skill. Such is the case for anthropogenic emissions, upon which climate projections made by the IPCC rest.
- The anthropogenic perturbation of CO_2 must be so small to lie almost within the noise of the calculation. It can represent but a few percent of increased CO_2. That, in turn, requires a removal time, which must be in the order of just a few years. Anthropogenic CO_2 is then removed from the atmosphere almost as fast as it is introduced,

sharply limiting its accumulation in the atmosphere (see the section on 'The relationship of carbon 14 to the removal of CO_2').

Physical mechanisms controlling greenhouse gas emissions

Closer inspection of the physical mechanisms, through which observed warming can produce the observed evolution of CO_2, shows that oceanic CO_2 emissions, E_O, depend primarily on wind speed and the atmosphere–ocean contrast of the respective mixing ratios r_O and r_A, while temperature induced emissions with a sensitivity $Q_O \approx 3\%/°C$ only contribute a minor fraction.

Different to the oceans, CO_2 emissions, E_L, over land follows from soil respiration and have a much higher temperature sensitivity, particularly over tropical land, with $Q_L > 30\%/°C$. This makes emissions from land determinative (Palmer et al. 2019). Therefore, despite the smaller land surface area, soil respiration can well be considered as the dominating temperature-dependent source of CO_2 (Salby & Harde 2022).

With a mean sensitivity $Q = \delta E/E/\delta T = 30\%/°C$ over tropical land and ocean (see Salby & Harde 2022), and an observed exponential response of the emissions $E(\delta T) = E(0) \cdot e^{Q \cdot \delta T}$, with δT as the observed SST anomaly (Kennedy et al. 2019), the measured net emissions (see Figure 9.3, blue line) can well be reproduced, now with a correlation of 80%. But different to the correlation deduced empirically in the previous section, this calculation even explains quantitatively the thermally induced emissions.

Unlike thermally induced emissions from the surface, anthropogenic emissions operate directly in the atmosphere. As discussed in the section 'Influence of tropical warming', the continuous re-supply of anomalous CO_2 in the atmosphere keeps the troposphere and surface layer out of equilibrium. Thereby, it limits re-emission and its offset of direct absorption, leaving effective absorption fast.

Integration of the total emissions, consisting of the thermally induced and anthropogenic contributions, and also including seasonal oscillations, gives the atmospheric mixing ratio r_A, which can directly be compared with observations. Figure 9.4 displays the measured CO_2 concentration at Mauna Loa (blue line) (CDIAC 2017). Superimposed

Figure 9.4 Calculated CO_2 mixing ratio r_A compared against the measurement at Mauna Loa

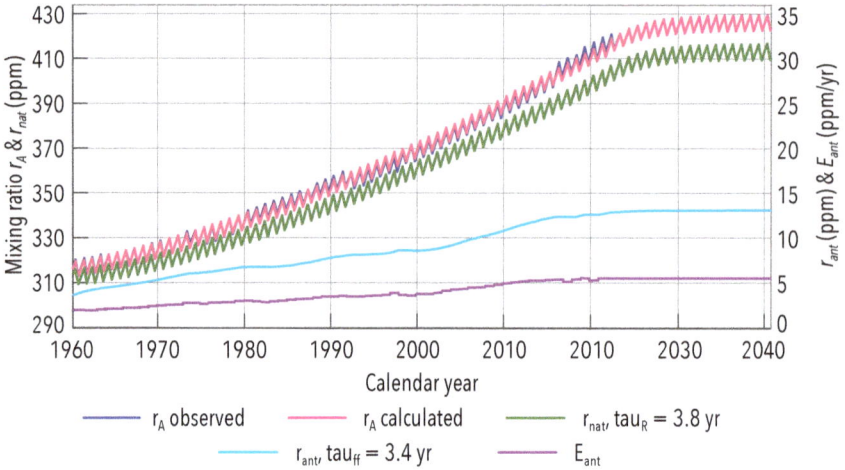

Source: Calculated CO_2 mixing ratio r_A (magenta), compared against measurement at Mauna Loa (blue) that lies mostly beneath calculation. Plotted separately is the native contribution r_{nat} (green), the anthropogenic fraction r_{ant} (aqua) and emission ratio E_{ant} (lilac) (right ordinate).

is the calculation (magenta line), which closely tracks the observed evolution of CO_2 and almost completely covers the observation.

Separately plotted is the native CO_2 fraction (green line) with a reference concentration in 1958 of $r_{nat}(1958) = 314$ ppm and on top the thermally induced component over the Mauna Loa Era. The effective absorption of the native emissions is regularly limited by the residence time, with $\tau_R \approx 3.8$ years (see IPCC 2021, Figure 5.12), but continuously compensated by equivalent emissions.

The anthropogenic contribution r_{ant} is shown as an aqua line, and the anthropogenic emission rate E_{ant} as a lilac line, both right ordinate. With a carbon mass ratio of the troposphere to surface layer of $m_T/m_S = \frac{1}{4}$, this reproduces the calculation in the second section with 13 ppm in 2022 (right ordinate) and corresponds to an effective absorption time of 2.4 years. For a lower mass ratio, the anthropogenic fraction and absorption time will even be smaller (Section 2; Salby & Harde 2022).

Figure 9.4 also reveals that for future constant emissions (human and natural), the concentration would come to equilibrium within less than one decade and only additionally increase by 9 ppm.

Like CO_2 emissions, CH_4 emissions also increase with temperature. It, too, is emitted by biomass, chiefly through anaerobic processes that operate in well-irrigated regions, such as wetlands; those influences magnify CH_4 emissions, particularly from tropical land areas, where biomass and precipitation are abundant.

The simultaneous intensification of CO_2 and CH_4 emissions is precisely what is expected from observed warming in the tropics. Therefore, this single physical mechanism provides a unified understanding of the joint increase of these greenhouse gases, one that follows naturally from thermally induced emissions.

Conclusion

An in-depth analysis reveals that thermally induced emissions in the tropics closely track observed net emissions of CO_2. It, therefore, accounts for the preponderance of CO_2 net emissions, which in turn determines anomalous CO_2. The strong correspondence to observed net emissions follows theoretically from their behaviour in the tropics – but not in the extra-tropics, where anthropogenic emissions are concentrated. Independently, the same correspondence to observed net emissions follows empirically from anomalous temperature in the tropics, as well as from a time-lag analysis of anomalous CO_2 (Humlum et al. 2013).

In both theoretical and empirical evaluations, thermally induced emissions of CO_2 represent interannual intensifications of net emissions, notably during episodes of El Niño. Represented equally well is the long-term intensification of net emissions during the last half century. The strong correspondence to observed changes indicates that, although operating on disparate time scales, both unsteady components of CO_2 net emissions share the same physical mechanism.

In relation to CO_2, what is responsible for that warming is immaterial. Its influence on CO_2 should not be confused through circular reasoning. The observed warming, which forces increased CO_2 through intensified net emissions, cannot itself be the result of increased CO_2. Otherwise, anomalous CO_2 and net emissions that force it would have increased twice as much as:

1. the increase required to produce the observed warming, plus
2. the thermally induced response to that warming, which, irrespective of what caused the warming, induces an intensification of net emission that is nearly identical to what is observed.

The direction of causation is also clear from the interdependence of net emissions and temperature, for interannual fluctuations as well as for the long-term increase. In addition to having strong coherence with temperature, the two unsteady components of net emissions have the same phase relationship to temperature, both varying *in phase* with temperature. The strong coherence and in-phase relationship to temperature reveal that, irrespective of time scale, changes in tropical temperatures induce simultaneous changes in CO_2 net emissions.

Under the opposite direction of causation, were the observed changes of tropical temperature induced by changes of CO_2, they would result in a fundamentally different phase relationship. The time scale of thermal damping, which drives temperature towards thermal equilibrium, is only a couple of weeks. It is much shorter than both unsteady time scales. Through anomalous radiative forcing, the comparatively gradual changes of CO_2 would, therefore, induce simultaneous changes of temperature, in phase with CO_2. However, net emission, which changes CO_2, must lead CO_2 by a quarter cycle. Net emissions would thus also lead temperature by a quarter cycle – behaviour contradicted by their observed in-phase relationship. Net emission of CO_2, which determines anomalous CO_2, is forced by changes of tropical temperature – not vice versa (see also Koutsoyiannis et al. 2023).

From the IPCC's own estimates of extraneous carbon reservoirs, the anthropogenic contribution to increased CO_2 has been shown to be no more than 15% to 35% (Harde & Salby 2021; Harde 2017; Harde 2019; Berry 2021; Harde 2023). These preceding considerations even show that this perturbation can be expected to contribute less than 10%. This is in strong contradiction to the IPCC's view, which presumes that 44% of the anthropogenic emissions are accumulating in the atmosphere as a so-called 'airborne fraction' and part of it will stay there for up to a few hundred thousand years (IPCC 2021, Chapter 5, Figure 5.7).

10 Natural Climate Drivers versus Anthropogenic Drivers

Associate Professor Antero Ollila

The IPCC's prevailing theory for global warming is that it is anthropogenic global warming (AGW), in which the major drivers are greenhouse gases (GHGs) from human causes. According to the IPCC, albedo, aerosols, and clouds have had cooling effects, and natural drivers have almost no role at all (<0.8%). This study explores the evidence to show that the role of natural influences, such as these, has been understated by the IPCC.

The strength of CO_2 as a greenhouse gas

Radiative forcing is a measure of the influence a factor has in altering the balance of incoming and outgoing energy in the Earth-atmosphere system. The effective radiative forcing (ERF) is the final radiative forcing (RF) at the top of atmosphere (TOA) for a particular forcing agent; according to the IPCC (2021) definition, it is the sum of the instantaneous radiative forcing (IRF) and the adjustments. The RF value caused by doubling the carbon dioxide (CO_2) concentration from 280 parts per million (ppm) to 560 ppm is shown as 2 × CO_2. In other words, if the atmospheric CO_2 concentrations doubled, the amount of thermal radiation escaping into space would be reduced by 3.9 Wm^{-2}. This figure is used to calculate climate sensitivity values as global surface temperature changes.

The 3.7 Wm^{-2} of 2 × CO_2 cited in AR5 (IPCC 2013) is from Myhre et al. (1998). It is based on spectral calculations at the tropopause (3.55 Wm^{-2}) and the stratospheric adjustments of 0.16 Wm^{-2}, which is

4.5% of the ERF value. In 2010, Schmidt et al. (2010) called this value a 'canonical estimate' as it seemed to be unchallenged.

In AR6, the IPCC writers, using a questionable rationale, introduced a higher figure for $2 \times CO_2$. In that paper, three $2 \times CO_2$ values were referred to. These were:

- Etminan et al. (2016) – 3.75 Wm^{-2}
- Meinshausen et al. (2020) – 3.75 Wm^{-2}
- Smith et al. (2018) – 3.70 Wm^{-2}.

Etminan et al. (2016) and Meinshausen et al. (2020) were both based on spectral calculations at the TOA, while Smith et al. (2018) was based on the simulation of eleven general circulation models (GCMs) applying the average IRF values at the TOA and the adjustments.

The IPCC's final ERF is now 3.9 Wm^{-2}. This is a combination of an IRF value of 2.6 Wm^{-2} plus stratospheric cooling of 1.1 Wm^{-2}, and the tropospheric adjustment of 0.2 Wm^{-2}. The IPCC (2021) does not pay any attention to these significant differences between AR5 and AR6.

There are also different $2 \times CO_2$ values, calculated by the line-by-line (LBL) method, which are not referred to in AR6 (IPCC 2021). The value of Barrett et al's (2006) is 3.1 Wm^{-2}, while that of Schildknecht (2020) is 3.0 Wm^{-2}. The Wijngaarden and Happer's (2020) $2 \times CO_2$ value is 3.0 Wm^{-2}. Harde (2013) also applied his own LBL calculations and his two-layer atmospheric model to arrive at a value of 2.4 Wm^{-2}. Miskolczi & Mlynczak (2004) carried out extensive LBL calculations with different atmospheric compositions, and their $2 \times CO_2$ value was 2.53 Wm^{-2}.

Ollila (2023a,b) has reported a $2 \times CO_2$ value of 2.65 Wm^{-2} using LBL calculations with the Spectral Calculator tool (GATS 2014) and by using the HITRAN (2021) database and water-continuum model. Ollila's calculations are in line with Ohmura (2001), who states that 98% of total LW absorption happens in the troposphere, and therefore the CO_2 absorption does not increase in the stratosphere, but it is saturated below the altitude of 1 km. The IRF value (the RF at the troposphere) of Smith et al. (2018) is 2.6 Wm^2, which is relatively close to the $2 \times CO_2$ values of Harde (2013) and Miskolczi and Mlynczak (2004), and almost the same as the value of Ollila (2023).

Positive water feedback or not

Positive water feedback is a cornerstone in any GCM, and in the simple model used by the IPCC. In the IPCC's AR5 (2013, p. 667) it states, 'Therefore, although CO_2 is the main control knob on climate, water vapour is a strong and fast feedback that amplifies any initial forcing by a typical factor between two and three.'

The theoretical justification of positive water feedback is based on the equation of Clausius–Clapeyron. This equation represents the pressure–temperature relationship in an atmosphere saturated with water vapour. The real atmosphere is not saturated by water vapour – the atmosphere's saturation is around 70% on average – and, therefore, the theoretical basis for this is weak. The direct humidity and temperature measurements from 1980 onwards show no positive water feedback over the longer term (see Figure 10.1).

Figure 10.1 The temperature trend and total precipitable water (TPW) trends from 1980 to 2020

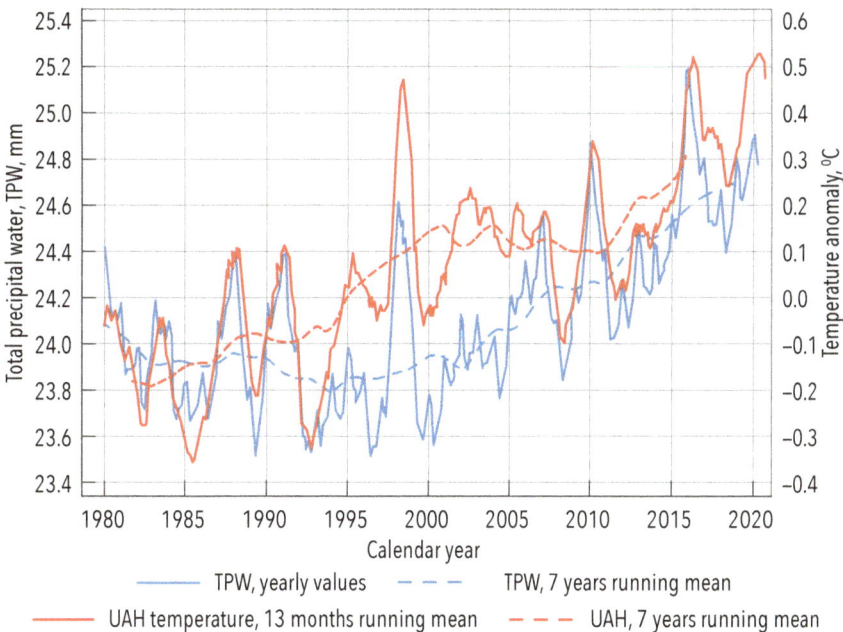

Source: Ollila, A 2021, 'Global Circulations Models (GCMs) simulate the current temperature only if the shortwave radiation anomaly of 2000s has been omitted'.

The long-term value of temperature (UAH 2022) has increased by about 0.4 °C from 1979 to 2000 but the TPW (NOAA 2021) values show a negative trend, which conflicts with the positive water feedback theory.

The surface temperature values can be calculated using a simple equation, as defined by the IPCC (2013, p. 664):

$$dTs = \lambda \times RF \qquad \text{(Equation 1)}$$

where dTs is the global mean surface temperature change (°C), and λ is the 'climate sensitivity parameter'. The IPCC (2001) reported that 'λ is the nearly invariant parameter (typically about 0.5 K/(Wm^{-2})'. This λ value was taken from the study of Ramanathan et al. (1985). The ERF value of 2.70 Wm^{-2} in AR6 results in a warming of 1.27 °C, meaning the λ value of 1.27 °C/2.70 Wm^{-2} = 0.47 °C/(Wm^{-2}).

The value of λ can be calculated from the total energy balance of the Earth by equalising the absorbed and the emitted radiation fluxes (Schlesinger and Ramankutty (1994); Ollila (2023). Using the average radiation CERES (2021) flux values for the period 2008–2014, $\lambda = 0.265$ K/(Wm^{-2}), Ollila (2021) has shown that by using this λ value, the temperature changes from 2012 to 2020 can be simulated with insignificant error.

Climate sensitivity

Climate sensitivity (CS) is a useful measure, telling us how much the Earth's surface temperature (Ts) would increase if the CO_2 concentration increased from 280 ppm to 560 ppm (2 × CO_2). There are two types of climate sensitivity, namely transient climate response (TCR), and equilibrium climate sensitivity (ECS) (IPCC 2013). According to AR6 (IPCC 2021) the 'TCR is a surface temperature response for the hypo-thetical scenario in which atmospheric carbon dioxide (CO_2) increases at 1% yr^{-1} from pre-industrial to the time of a doubling of atmospheric CO_2 concentration (year 70)'.

IPCC (2013, p. 1112) states that 'TCR is a more informative indicator of future climate than ECS'. Applying Equation 1 gives the average TCR value of 1.83 °C (= 0.47 °C/(Wm^{-2}) *3.90 Wm^{-2}), while the best estimate of AR6 (IPCC 2021) is 1.8 °C with the likely range from 1.4 °C to 2.2 °C.

The literature survey of non-IPCC TCR values can be divided into three major categories based on the research method, namely:

A. using the $2 \times CO_2$ values of the IPCC.
B. using the $2 \times CO_2$ by applying researchers' own LBL analysis calculations.
C. using observed surface temperature values and other climate data.

The TCR values of category A vary from 1.15 °C from Bengtson and Schwartz (2013), 1.2 °C from Schlesinger and Ramankutty (1994), and 1.33 °C from Lewis and Curry (2015). These results mean there is no positive water feedback. These results are fully in line with the IPCC (2007), that 'with no feedback operating, the global warming from GCMs would be around 1.2 °C'.

The results of category B hardly vary: 0.6 °C by Barrett et al. (2006); 0.48 °C by Miskolczi and Mlynczak (2004); 0.51 °C by Ollila (2012); 0.7 °C by Ollila (2023); 0.5–0.7 °C by Kissin (2015); 0.7 °C by Harde (2017); and 0.5 °C by Schildknecht (2020). The only explanation is that these researchers have found a smaller $2 \times CO_2$ value than 3.7–3.9 Wm^{-2} and they have not found positive water feedback.

The survey of research studies of category C reveals that the TCR values vary from 0.0 °C of Fleming (2018) to 1.2 °C of Otto et al. (2013). The results of this category are not reliable since they are too heavily dependent on other climate drivers like solar irradiation variations, volcanic impacts, surface albedo changes, and so on (Kissin 2015). Usually, the elimination of these effects has not been considered at all.

Carbon dioxide circulation and time delays in the atmosphere

The only way to identify this behaviour is to build a model simulating the carbon cycle between the atmosphere, the ocean, and the land. About 25% of the atmospheric CO_2 changes every year because the oceans absolve and dissolve CO_2, and in the same way land plants photosynthesise and respire CO_2.

In AR6, the IPCC (2021) states: 'Over the past six decades, the average fraction of anthropogenic CO_2 emissions that has accumulated in the atmosphere (referred to as the airborne fraction) has remained nearly constant at approximately 44%.'

The IPCC claims that all the increased CO_2 amount of 279 gigaton carbon (GtC) in the atmosphere in 2019 is entirely anthropogenic. These figures mean that the yearly sequestration rate of the anthropogenic CO_2 emissions would be 56%. Since 25% of the atmospheric CO_2 changes every year, it would mean that the sinks would strongly prefer anthropogenic CO_2. Nature – or even man-made processes – cannot do it since the difference in composition is only 0.022%.

The present atmospheric anthropogenic CO_2 amount is only 70–71 GtC according to Ollila (2020), and Berry (2021).

The nuclear bomb tests in the atmosphere between 1945 and 1964 accidentally created a tracer test situation for carbon-14 (^{14}C). The residence time of sixteen years gives an excellent fit, meaning a relaxation time of 64 years. This tracer test by the ^{14}C corresponds perfectly to the behaviour of anthropogenic CO_2.

In AR5 the IPCC (2013, p. 469) writes: 'The removal of human-emitted CO_2 from the atmosphere by natural processes will take a few hundred thousand years (high confidence).'

This result is in direct conflict with the tracer test results with radiocarbon.

The decadal climate oscillations during the Holocene

60- to 88-year periodicities have been found in regional and global measurements and proxies over thousands of years, similar to those of global temperature (HadCRUT4 2021), Indian monsoons (Agnihotri et al. 2002), northeast Pacific coast sediments (Patterson et al. 2004), cosmogenic isotope concentrations of ^{14}C and beryllium-10 (^{10}Be) (Attolini et al. 1987; Cini Castagnoli et al. 1992), auroral records (Feynman & Fougere 1984), and tree-ring analyses (Lin et al. 1975; Peristykh & Damon 2003; Ollila & Timonen 2023).

Gleissberg (1958) found that the solar cycles weaken and strengthen in an 80- to 90-year cycle. Therefore, the periodicity of the 88-year Gleissberg cycle is related to the eleven-year Schwabe (1843) cycle. Scafetta (2010) and Ollila (2017) have introduced and analysed the planetary oscillation called astronomic harmonic resonances (AHR), which creates a 60-year oscillation pattern. The 60-year oscillation is called the Atlantic multidecadal oscillation (AMO) by Kerr (2000). Chen et al. (2018) identified the same oscillation in the Pacific, and it was termed the Pacific multidecadal oscillation (PMO).

Ollila and Timonen (2023) analysed the year-accurate tree-ring data series called the Finnish Timberline Pine Chronology (FTPC 2022) from the year 1000 onwards. They found that the tree-ring variations can be explained with two oscillation periods of 60- and 88-years. The 60-year period matches the AHR phenomenon, and the longer oscillation is a well-known 88-year Gleissberg oscillation (see Figure 10.2).

Figure 10.2 FTPC 31-year running mean signal and combined AHR and Gleissberg signal

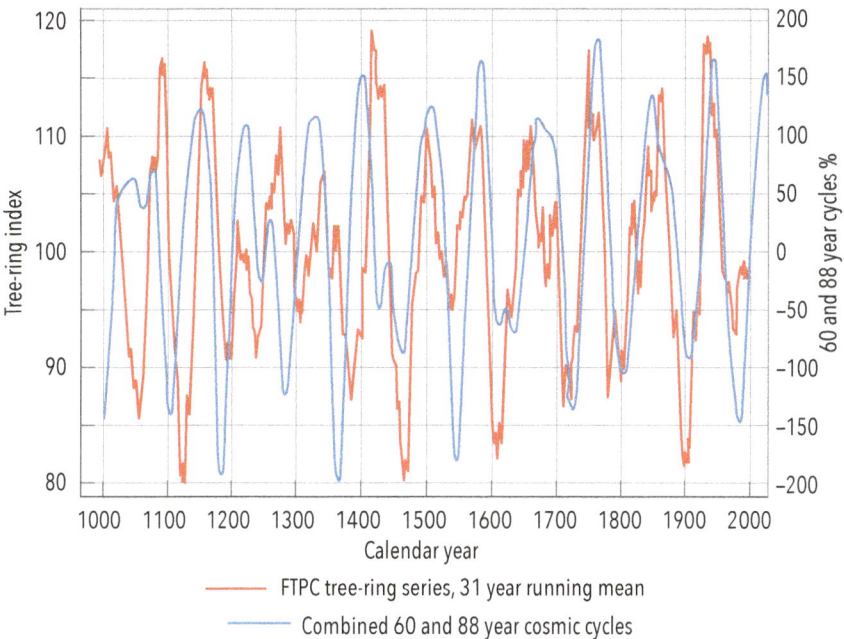

Source: Ollila, A & Timonen, M 2023, 'Two main temperature periodicities related to planetary and solar activity oscillations' HAL Open Science. The 60- to 88-year oscillations had their simultaneous maximums in 1940. When the oscillation phases changed to negative phases, the cooling effect of the 60- and 88-year oscillations became dominant over the greenhouse gas effects and caused global temperatures to decrease. This happened from 1940 to 1975. Similarly, when the 60- and 88 oscillations turned from a negative to a positive phase, global warming accelerated, as it did after 1975, finally increasing the global temperature by about 0.25 °C up until 2000. The IPCC did not recognise this temperature behaviour in its temperature reconstruction during the 1900s.

Century- and millennial-scale climate oscillations

Century- and millennial-oscillations have a major role in explaining long-term variations during the anthropogenic period from 1750 to the present. If there are longer periodicities than about 250 years, then these would be possible explanations for the present-time warming, at least partially.

Century- and near-millennium scale research studies have typically used ice-core drilling samples of Antarctica and Greenland; the other group of studies has used cosmogenic analyses of ^{14}C and ^{10}Be samples from other sources like marine and lake sediment records, and delta-O-18 ($\delta^{18}O$) (the ratio of stable isotopes oxygen-18 (18O) and oxygen-16 (16O)) records of speleothems (geological formations of mineral deposits in natural caves).

Because the temperature and CO_2 variations are smaller in Antarctica than in Greenland, the global oscillations should also easily be found in the Northern Hemisphere. The periodicities of Greenland's ice-core records according to Vinther et al. (2010) have been 1270, 1470, and 2550 years. In a later article of Vinther's (2011), a dominant period is about 1000 years peaking at 1000 and 2000 years. Bond (1997) has found the same 1470 ± 500 years periodicity in the North Atlantic Sea sediments during the Holocene.

The Sun's activity variations

The Earth receives about 99.97% of its energy from the Sun. The Sun's radiated energy measure is total solar irradiance (TSI), which has both long-term variations in the millennium scale and short-term variations, such as the eleven-year cycle of Schwabe (1843). Solar magnetic field variations are responsible for solar irradiation changes. There are two main categories of methods used in evaluating historical TSI values: sunspot records starting from 1610; and cosmogenic isotopes of ^{10}Be and ^{14}C applicable for millennial periods.

Hoyt and Schatten (1993), Lean (1995; 2004; 2010), and Bard et al. (2000) have applied different methods and their results are shown in Figure 10.3.

Figure 10.3 TSI reconstructions of Lean (2004), Hoyt and Schatten (1993), and Bard et al. (2000)

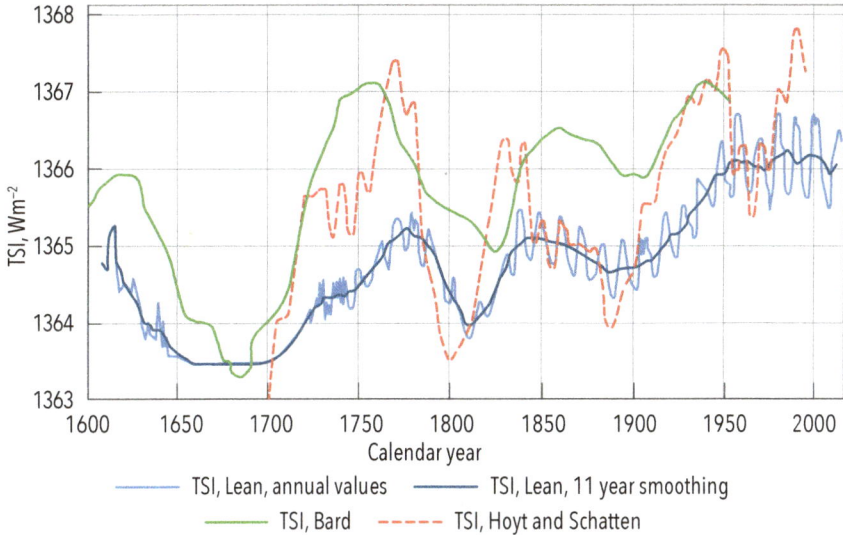

Source: Ollila, A 2017, 'Semi empirical model of global warming including cosmic forces, greenhouse gases, and volcanic eruptions', *Physical Science International Journal.*

The TSI reconstruction of Velasco Herrera et al. (2015) shows the same general TSI trends as above and they also predict that the TSI trend has a minimum of around the 2050s. Connolly et al. (2021) have carried out a comprehensive review study of the Sun's effects on the Northern Hemisphere temperature trends. The results show a common feature in all TSI reconstruction studies, which is that in around 1900 the TSI value was about -2 Wm^{-2} lower; in the 1930s about $+1$ Wm^{-2} higher; and from 1990 onward about 1.5 Wm^{-2} higher than the reference level.

Kauppinen et al. (2014) and Ollila (2013) have found from satellite cloudiness observations that a 1% cloudiness change causes a 0.1 °C temperature change. Ollila (2017) has suggested that the TSI impacts cause cloudiness changes and due to this effect, an RF change should be multiplied by a factor of 4.2. By applying this factor, the temperature impact of the TSI change of 1.1 Wm^{-2} from 1750 to 2020 would be 0.32 °C.

The varying magnitudes of climate drivers have been summarised in Table 10.1.

Table 10.1 The main anthropogenic and natural drivers of surface temperature changes according to IPCC (2013), IPCC (2021), and the review study of Ollila (2023) called Natural Anthropogenic Global Warming (NAGW) from 1750 to 2019. The values in parentheses are calculated according to the IPCC science if the shortwave (SW) anomaly of the 2000s is included (Ollila 2021).

Driver	IPCC AR5, °C	IPCC AR6, °C	NAGW, °C
Carbon dioxide	0.84	1.01	0.36
Methane	0.49	0.28	0.14
Nitrogen oxide	0.09	0.10	0.04
Other anthropogenic gases	0.18	0.44	-
Greenhouse gases	1.59	1.83	0.54
Albedo, volcanic	−0.08	−0.09	-
Aerosols, clouds, and contrails	−0.42	−0.49	-
Anthropogenic totally	1.11	1.28	0.54
Solar	0.03	−0.01	0.32
SW radiation anomaly	-	0.00 (0.78)	0.43
Drivers totally	1.17	1.27 (2.03)	1.29
Observed temperature change	0.85	1.29	1.29
Deviation	+37.7%	−1.6% (57%)	0.0%

Source: Ollila, A 2023, 'Natural climate drivers dominate in the current warming', *Science of Climate Change.*

The most important result is that according to the AR6 (IPCC 2021), the contribution of CO_2 during the industrial era has been 1.01 °C, but according to this study it is 0.36 °C, and according to Harde (2022) it is 0.34 °C.

The trend in climate driver magnitudes from AR5 to AR6 is consistent. The most striking feature is the temperature deviation percentage change from +37.7% in 2011 to −1.6% in 2019 (material years of AR5 and AR6). This change cannot be explained by the abrupt increase of anthropogenic drivers as noted in Table 10.1. The most probable reason is the emerging SW radiation anomaly resulting in +0.43 °C from 2001 to 2019, as indicated in the last column.

Conclusion

In the AR6, the overall radiative forcings (RF) were 2.70 Wm^{-2} in 2019, and the temperature increase was 1.27 °C. Research studies are showing significantly lower RF and climate sensitivity values for anthropogenic climate drivers. Natural climate drivers have been suggested as the partial or total solution for global warming, including solar radiation changes, cosmic forces, and multidecadal, and century- and millennial-scale climate oscillations. The cloudiness changes seem to have a significant role in magnifying cosmic effects like the TSI changes. The AMO and the Gleissberg cycle can explain the ups and downs of the global temperature in the 1900s, which happened in 60- and 88-year periods. Century-scale oscillations have been identified in many research studies. The IPCC science opponents have revealed crucial errors in the carbon cycle models of the IPCC, essentially demonstrating smaller atmospheric anthropogenic CO_2 amounts and relaxation times. The summary of natural climate drivers together with anthropogenic drivers constitutes an alternative theory called Natural Anthropogenic Global Warming, in which natural drivers have a major role in dominating the warming during the current warm period. These results mean that there is no climate crisis, nor any need for prompt CO_2 reduction programs.

11 Atmospheric CO_2 – Natural and Anthropogenic Contributions of the Past Half-century

Dr Tom Quirk and Dr Michael Asten

Summary

Differing characteristics in isotopes of carbon allow us to differentiate between carbon dioxide (CO_2) originating from plant material (including fossil fuels) and from the ocean. Understanding and quantifying the contributions of CO_2 from plants and the ocean into the atmosphere is important in distinguishing the relative human (anthropogenic) versus natural contributions.

Plants sequester carbon through photosynthesis. During this process, plants take in atmospheric CO_2 and convert it into organic carbon compounds, storing carbon in their biomass. Respiration by plants releases some of this stored carbon back into the atmosphere as CO_2.

The world's oceans are the largest reservoir of carbon on Earth. They absorb a significant amount of atmospheric CO_2 through a process called dissolution. The solubility of CO_2 in seawater depends on factors like temperature and pressure. Colder waters and higher pressures increase the ocean's capacity to dissolve CO_2. While oceans act as a major carbon sink, they also release CO_2 through a process known as degassing. The oceans' CO_2 emissions can be influenced by changes in ocean circulation patterns, temperature, and biological productivity.

The CO_2 released by plants into the atmosphere and the CO_2 released by the oceans into the atmosphere have distinctive isotopic signatures. Plants, for example, preferentially take up the lighter isotope of carbon (^{12}C) during photosynthesis. It is not possible to distinguish the plant component of CO$_2$ as released for example by rotting of plant material, from that released by the burning of fossil fuel, but it is possible to distinguish the ocean component that is often confounded with human emissions.

This summary highlights key results presented by Quirk and Asten (2023). The analysis uses both atmospheric CO$_2$ concentrations and the accompanying isotopic measurements of CO$_2$ (expressed as $\delta^{13}C$) over a period of 40 years from 1978 to 2015, observed at ten different latitudes between 90°S to 82°N. Atmospheric CO$_2$ is separated into two components of CO$_2$ attributable to deep ocean and to plant (including fossil fuel) sources. The isotopic values assigned to the two components are $\delta^{13}C$ = 0‰ and –26‰, respectively.

The latitude variations in residual source component CO$_2$ show the ocean source component peaking at the Equator. This contrasts with the residual plant source component that peaks in the Arctic Circle region. Seasonal comparisons show no change in the ocean component peaking at the Equator and no significant changes in its variation with latitude, while the plant component shows seasonal changes of the order of 15 parts per million (ppm) at high latitudes. The ocean component shows clear anomalous behaviour in the three years following the 1989 Pacific Ocean Regime Shift (a shift independently identified from the changed biological time series). By contrast, the residual plant component shows a correlation in the timing of maxima in its annual variations with the timing of El Niño events from 1985 to 2015. It also shows a discontinuity in annual variation coinciding with the 1995 Atlantic Multi-decadal Oscillation (AMO) phase change.

We conclude that the ocean and plant components of atmospheric CO$_2$ relate to independent sources of atmospheric CO$_2$ and have approximately equal magnitudes. The observations are consistent with a hypothesis that variations in the ocean components have an origin from upwelling water from deep ocean currents, and variations in plant components are dominated by a combination of fossil fuel CO$_2$, phytoplankton productivity, and forest and peat fires, which primarily occur in the Northern Hemisphere.

Introduction

It has been previously shown that the apparent smooth and continuous rise in atmospheric CO_2 concentrations can be broken into a series of trend changes associated with ocean decadal phase changes (Quirk 2012). We extend that association through the separation of CO_2 concentrations on the basis of carbon isotopes in order to distinguish between variations in CO_2 concentrations associated with Pacific Ocean regime changes, and variations associated with sea-surface temperature as represented in the Niño 3.4 index (NOAA n.d.).

We use the isotopic composition of CO_2 as observed at ten globally distributed stations in order to obtain estimates of annual and decadal variations in atmospheric CO_2 attributable to plant origins (both natural and anthropogenic processes) and to CO_2 transfer into the atmosphere from upwelling deep ocean waters.

The measure of the ratio of ^{12}C to ^{13}C found in a sample of CO_2 is expressed as $\delta^{13}C$, where the $^{13}C/^{12}C$ abundance ratio is of the order of 0.01. Variations in $\delta^{13}C$ are attributable to variations in the molecular weight of CO_2, which affect the rate of chemical reactions and photosynthesis.

There are now about 60 years of atmospheric CO_2 concentration measurements available from the Scripps Institution of Oceanography (SIO) (Keeling et al. 2001) and 40 years of $\delta^{13}C$ isotope measurements of CO_2 commencing in 1978. The SIO stations and years with both CO_2 and $\delta^{13}C$ measurements are listed in Table 11.1. The additional column shows the percentage of months with no data from 1986 to 2015.

We selected from 1986 to 2015, the span of years with the most complete record and avoided the 32-month period from 2016 to 2018 on Christmas Island, for which there was no data.

This analysis uses the CO_2 concentration and isotope measurements after 1978 to divide the atmospheric CO_2 concentration into an 'ocean' and a 'plant' component by assuming values for the isotopic composition for the two components.

The purpose of this study is to demonstrate a means of computing the fraction of gaseous CO_2 in atmospheric samples that can be attributed to a plant origin.

Table 11.1 Scripps Institution of Oceanography stations and years for CO$_2$ and δ^{13}C measurements

	Latitude	Longitude	Elevation (metres)	Years with CO$_2$ and δ^{13}C	% Months with no Data	
					All	1986-2015
Alert	82° N	63° W	210	1986-2018	4.55%	4.72%
Point Barrow	71° N	157° W	11	1983-2018	4.17%	4.72%
La Jolla	33° N	117° W	10	1979-2018	14.38%	10.28%
Kumukahi	19° N	155° W	3	1981-2018	1.10%	0.28%
Mauna Loa	19° N	156° W	3397	1981-2018	2.19%	0.00%
Christmas Island	2° N	157° W	2	1978-2018	32.11%	26.67%
American Samoa	14° S	171° W	30	1985-2018	3.92%	3.33%
Kermadec	29° S	178° W	2	1985-2018	53.43%	51.94%
Baring Head	41° S	175° E	85	1986-2018	40.91%	39.17%
South Pole	90° S		2810	1978-2018	7.52%	3.06%

Source: Keeling et al. 2001. *Note that Christmas Island is part of the Republic of Kiribati in the Pacific Ocean.

The isotopic approach used in this paper allows a phenomenological study of atmospheric CO$_2$ origins from upwelling deep ocean sources, and from plant sources – including both anthropogenic and natural sources that release CO$_2$ into the atmosphere. This approach does not require assumptions or estimations of the dissolved inorganic carbon content of the ocean surface waters, nor does it quantify the uptake by the oceans of atmospheric CO$_2$. It does, however, provide quantitative data on seasonal, annual and latitudinal variations in fractions of atmospheric CO$_2$ attributable to ocean and plant sources.

Method, model and data

Deep ocean and plant contributions to CO$_2$

The basis of this analysis is that contributions to CO$_2$ derived from deep ocean and from plant sources take discrete and different values. For SIO δ^{13}C data, the standard ratio $R(^{13}C/^{12}C)_{std}$ is from the Vienna Peedee Belemnite (VPDB) isotope (Hoffman & Rasmussen 2022) where $R_{std} = 0.011180$.

Deviations in the isotope ratio as measured in air samples are expressed as

$$\delta^{13}C_{meas} = ((R_{meas}/R_{std}) - 1) \times 1000$$

For this analysis, we use two reference points as follows. For deep oceans:

$$^{13}C_{ocean} = 0‰$$

(Kroopnick 1985; Ohmoto 1986; Feely et al. 2008), and for plants:

$$\delta^{13}C_{plant} = -26‰$$

The plant value is representative of C3[1] plants (NOAA, The global conveyor belt); phytoplankton for north and south, high latitudes (O'Leary 1988); and fossil fuels (Goericke & Fry 1994).

The separation of deep ocean and plant contributions can be quantified using the two relations defined in Equations (1) and (2):

$$A_{meas} = A_{ocean} + A_{plant} \tag{1}$$

$$\delta^{13}C_{meas} = (\delta^{13}C_{ocean} \times A_{ocean} + \delta^{13}C_{plant} \times A_{plant})/A_{meas} \tag{2}$$

where:

- A_{meas} is the measured atmospheric concentration of CO_2 in ppm
- A_{ocean} is the component of deep ocean origin, in ppm
- A_{plant} is the component of plant origin, in ppm
- $\delta^{13}C_{meas}$ is the isotopic value of measured atmospheric CO_2
- $\delta^{13}C_{ocean}$ is the isotopic value of 'ocean' component atmospheric CO_2
- $\delta^{13}C_{plant}$ is the isotopic value of 'plant' component atmospheric CO_2.

So, the separate components calculated using the measured values for atmospheric CO_2, and assuming isotopic values for 'deep ocean' and 'plant' components, are given by Equations (3) and (4) as follows.

$$A_{ocean} = A_{meas} \times (\delta^{13}C_{meas} - \delta^{13}C_{plant})/(\delta^{13}C_{ocean} - \delta^{13}C_{plant}) \tag{3}$$

and

$$A_{plant} = A_{meas} - A_{ocean} \tag{4}$$

---

1   Most plants and crops are C3 plants; the first carbon compound produced during photosynthesis contains three carbon atoms.

Measured atmospheric concentrations are recorded in ppm by volume. Thus, separating an ocean and a plant source component requires converting ppm by volume to ppm by mass.

So,

$$\delta^{13}C = \{R(^{13}C/^{12}C)_{meas}/R(^{13}C/^{12}C)_{std} - 1\} \times 1000 \qquad (5)$$

and

$$R(^{13}C/^{12}C)_{meas} = (1 + \delta^{13}C/1000) \times R(^{13}C/^{12}C)_{std} \qquad (6)$$

where:

- $R(^{13}C/^{12}C)_{meas}$ is the ratio for the sample carbon-13 to carbon-12
- $R(^{13}C/^{12}C)_{std}$ is the ratio for the standard carbon-13 to carbon-12

Then, the carbon mass may be calculated in atomic mass units as

$$\text{Carbon} = \{12 + 13 \times R(^{13}C/^{12}C)_{meas}\}/(1 + R(^{13}C/^{12}C)_{meas}) \qquad (7)$$

So, the mass of CO_2 can be calculated, hence the conversion of ppm by volume to ppm by mass.

The mass correction is small, of the order of 0.0005 ppm compared with measurement errors of the order of 0.1 ppm. Although not significant, these mass corrections are applied to this data.

For the remainder of this paper, we use the term 'ocean component' and 'ocean source' to refer to that fraction of atmospheric CO_2 having the isotopic signature of deep ocean water.

Ocean and plant source component analysis for monthly values

The results of the monthly measurements of atmospheric CO_2 concentrations for the South Pole and Christmas Island from 1978 to 2015 and for Point Barrow from 1983 to 2015 are shown as the ocean component (ppm) in Figure 11.1 and the plant component (ppm) in Figure 11.2.

The seasonal variations seen in the measured CO_2 concentrations for the South Pole, Christmas Island and Point Barrow are found in the respective plant components but not to any significant extent in the corresponding ocean components.

Figure 11.1 Monthly ocean-component CO_2 concentrations at the South Pole, Christmas Island and Point Barrow for 1978 to 2015

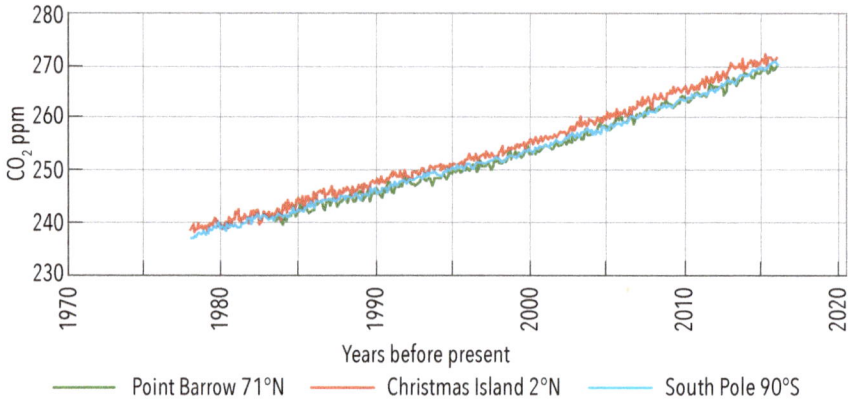

Source: Quirk and Asten 2023.

Figure 11.2 Monthly plant-component CO_2 concentrations at the South Pole, Christmas Island and Point Barrow for 1978 to 2015

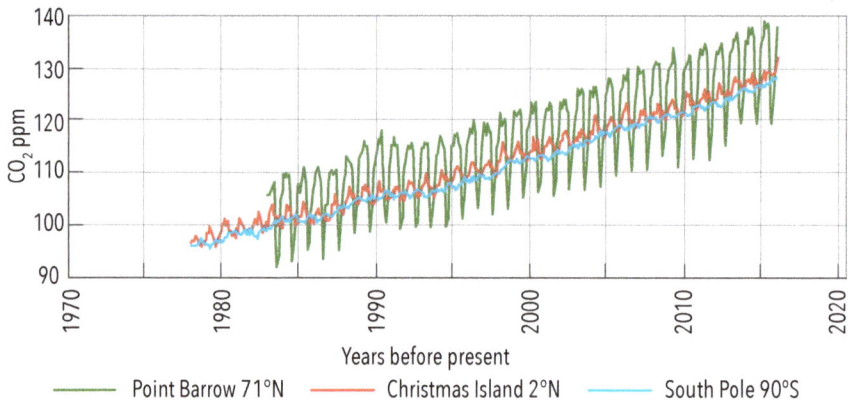

Note: an increase is evident around 1989 (the 'bubble').

Source: Quirk and Asten 2023.

There are two observations worth noting from this analysis:

1. A subtle increase, or 'bubble' above the trend in the measured CO_2 concentrations occurs from around 1988 to 1991. This feature overlaps in time the 1989 Regime Shift described by Hare and Mantua (2000), and is clearly evident in biological records, although

only weakly shown in indices of the Pacific climate. Consistent with that, we see that the 'bubble' is present in the plant component of atmospheric CO$_2$ as deduced from the isotopic analysis (Figure 11.2). However, the 'bubble' is not evident in Figure 11.1, which shows the equivalent plot of the ocean component of atmospheric CO$_2$.

2. For the ocean components shown in Figure 11.1, Christmas Island (2°N) near the Equator shows higher values for the entire time span of 1978 to 2015, while Point Barrow (71°N) in the Arctic shows the lowest values.

The average annual increases for each of the stations, computed for the time span from 1986 to 2015, are found through a least-squares fit to the measurements. An interesting result is that, for each site, the ocean and plant average annual increases are similar to within a few per cent.

We discuss the significance of these observations below; a full analysis is available at Quirk and Asten (2023).

Ocean and plant source component latitude analysis

Annual values
The behaviour of the ocean and plant components are further explored through analysis of each of the ten SIO latitudes listed in Table 11.1.

The variation in atmospheric CO$_2$ measurements shows increasing concentrations from the lowest values at the South Pole 90°S to a peak at Point Barrow 71°N, with residual annual variations of total CO$_2$ – less the value at the South Pole – for each station, averaged over the period 1986 to 2015.

The latitude variations in the residual annual variations of total CO$_2$ concentrations are quite different when separated into ocean and plant components. The component values, standard deviations, and errors of the mean are shown in Figure 11.3. The ocean component peaks near the Equator; however, the plant component peaks at the Arctic Circle latitudes of Point Barrow 71°N and Alert 82°N.

The analysis also shows significantly different residual values at Mauna Loa at an elevation of 3397 m compared to Kumukahi at the

Figure 11.3 Residual annual variations in CO_2 source components, less the value at the South Pole, averaged for 1986 to 2015

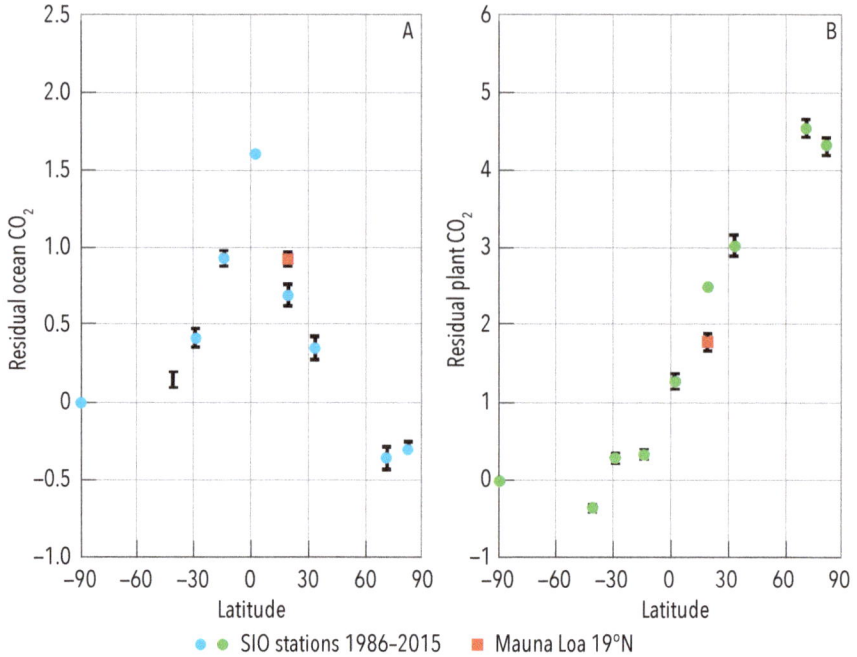

(a) The ocean component and (b) the plant component

Source: Quirk and Asten 2023.

same latitude but at an elevation of 3 m. The ocean-component residual at Mauna Loa is significantly greater than the Kumukahi residual, with this difference of residuals reversed for the corresponding plant components. These differences may be due to the elevation at which Mauna Loa measurements are made, which is above the sub-tropical inversion. This effectively means that the atmosphere is locally more mixed with contributions from the equatorial region. This would have the effect of increasing the residual ocean component relative to that found at Kumukahi at an elevation of 3 m. Likewise, the residual plant-component values at the elevated Mauna Loa site are mixed to a lower value than the Kumukahi residual plant value.

Latitude analysis for seasonal values

If the ocean source component is independent of the plant source component, then the ocean source peaking at the Equator should be independent of seasonal variations.

We found there is little change in the ocean components for the seasonally separated values relative to the residual annual averages, while the residual seasonal variations in the plant component of CO$_2$ are quite different. Plant components show changes of up to 16 ppm from the end of the boreal winter (February to March) to end of summer (August to September) at the far-north stations of Point Barrow and Alert.

The hypothesis is that these seasonal differences in the Northern Hemisphere are due to the influence on the carbon cycle of the boreal forests. The trees start to grow in spring, drawing CO$_2$ from the atmosphere until autumn when the end of summer growth reduces the draw-down of CO$_2$. In addition, it is expected that decaying plants return CO$_2$ to the atmosphere.

For the ocean component, the seasonal differences are not statistically different with the exception of Christmas Island 2°N, American Samoa 14°S and Kermadec Island 29°S. The peak in the ocean component for the boreal winter is shifted about 15°S. This may be explained by strong wind systems, particularly the trade winds, being comparatively enriched in the ocean component.

The plant component also shows evidence of atmospheric transport from the Northern Hemisphere to the Southern Hemisphere during the boreal winter, seen at latitudes 15°S to 30°S.

Analysis shows that at Point Barrow, 71°N, and at Alert, 82°N, the ocean component varies only slightly by 0.3 ± 0.2 ppm and 0.1 ± 0.1 ppm from boreal winter to summer, whereas there are large variations of 16 ppm and 15 ppm, respectively, for the plant component. At lower latitudes in the Northern Hemisphere, there are reduced differences from winter to summer. The spatial and seasonal distribution in ocean and plant source components differs by up to a factor of 50, and leads us to the conclusion that the ocean source of CO$_2$ is an independent source of atmospheric CO$_2$ that is decoupled from the plant source component.

Sensitivity of latitude analysis to assumptions of isotopic composition

The separation of observations of atmospheric CO_2 content into ocean and plant source components is made using assumptions that the CO_2 ocean component $\delta^{13}C$ is 0‰ and the plant component $\delta^{13}C$ is –26‰. The sensitivity to this choice is assessed by independently varying each component $\delta^{13}C$ by ± 2.

The perturbed values of $\delta^{13}C$ ocean give a maximum component variation of 0.1 ppm, while the corresponding perturbations in $\delta^{13}C$ plant give a maximum component variation of 0.4 ppm.

These perturbation studies demonstrate that the patterns of ocean and plant component changes with latitude are not significantly affected by variations of ±2‰ in the assumed baseline values used for $\delta^{13}C$. More details are given in Quirk and Asten (2023).

Annual differences of ocean and plant CO_2 components

Our results show that there is more annually averaged variability in the plant components compared to the ocean components. The simplest demonstration of variability comes from the standard deviations for the year-on-year changes, with the ocean components consistently less variable than the plant components. This may be attributable to the fact that the ocean dynamics change more slowly than the atmospheric dynamics that affect plant components.

The quantitative annual increases for atmospheric CO_2 and components from 1986 to 2015 show that the ocean component and plant components are approximately equal at 50% of the annual CO_2 concentration increase. More details are given in Quirk and Asten (2023).

Trend breaks in annual differences for the ocean component

The annually averaged CO_2 concentration shows a break point in its trend at year 2001 ±2. Graphical details in Quirk and Asten (2023) show there is a break in the ocean component at all latitudes, where the average ocean component of CO_2 increases from 0.731 ± 0.008 ppm per year for 1986 to 2000 to 1.066 ± 0.005 per year for 2002 to 2015.

The observed break in ocean-component CO$_2$ after 2001 is coincident with an inflection in global lower troposphere temperatures as measured by satellites (Earth System Science Center 2022). Tropospheric temperatures show a rising trend of the order of 0.15 °C per decade before 2001, and a temperature plateau starting from 2000 to 2002, extending to approximately 2015 (the duration of this study). Details are given in Quirk and Asten (2023, figures 11 and 12).

Likewise, an analysis of the global carbon cycle by Gruber et al. (2023) also shows a change in trend following the year of 2000. That analysis found an increase in CO$_2$ absorbed into the ocean from about 2000 to 2019.

In contrast to Gruber et al., the Earth System Science Center (2022), based on observational data only, demonstrated a change in the trend of movement of CO$_2$ from ocean to atmosphere.

A hypothesis, consistent with the observed break in the trends of atmospheric ocean-component CO$_2$ and tropospheric temperature around 2001, is that there is an increased upwelling of cold water in the oceans, which may then reduce global temperature and release CO$_2$ with an increased ocean-component isotopic mix into the atmosphere.

Annual differences in the plant-component CO$_2$; correlation with a phase change in major ocean current patterns

The plant-component CO$_2$ also shows an observed temporal change associated with ocean and atmospheric temperature shifts, including the aforementioned 'bubble' for the years 1988 to 1992, shown in Figure 11.2. The timing of this feature is also associated with the 1989 Regime Shift described by Hare and Mantua (2000). Replotting the data after trend removal shows not only the 'bubble' but a correlation of plant-component CO$_2$ with increasing latitude – which is a regional rather than global change (Quirk & Asten 2023, figure 14). The residual plant CO$_2$ values show increases in a northerly direction from the South Pole to Point Barrow at 71°N. The obvious hypothesis to explain this variation is that the source of atmospheric plant CO$_2$ is a combination of fossil fuel CO$_2$, phytoplankton productivity in the oceans, and forest and peat fires, which occur primarily in the Northern Hemisphere.

A further association of variations in plant-component CO_2 with ocean current variations is evident in the plot referenced above. There is a clear inflection in the residual trends of the annual change in plant-component CO_2 near the year of 1995. This coincides with a phase change in the Atlantic Multi-decadal Oscillation (AMO) from negative to positive, as described by Alexander et al. (2014). That phase change is also associated with synchronous changes in numbers of small pelagic fishes (Alheit et al. 2014). These observational changes are consistent with an interpretation of decreased phytoplankton productivity after 1995, which in turn may explain the observed decrease in numbers of small pelagic fishes. The pelagic fish time series shows an increase from 1995 to 2010, implying an increase in phytoplankton productivity. This would cause the increased removal of CO_2 in the shallow ocean; this outcome, in turn, is associated with a decrease in the plant component of atmospheric CO_2. Thus, variable ocean sinks for CO_2 are directly coupled to the atmosphere as the concentration and isotopic composition of the atmosphere are varied.

Correlation of global plant-component CO_2 variations with the El Niño Southern Oscillation

It is possible to estimate the annual changes in global atmospheric CO_2 concentrations and the separate ocean and plant CO_2 components using the nine surface stations listed in Table 11.1 (excluding Mauna Loa because of its altitude). The nine stations are taken to represent values in latitude bands defined by the average of the station latitudes, with the exception of the latitude extremes represented by Alert and the South Pole. The band latitudes are then used to calculate the fraction of the global surface within the bands. The plotted results are shown in Figure 11.4, below.

The annual changes in global atmospheric CO_2 concentrations and annual changes in global plant and ocean CO_2 components, when compared to the El Niño Southern Oscillation (ENSO) variation, show strong 1997 to 1998, and 2016 to 2017, El Niño events. Between 1988 and 2013, there are ten El Niño peaks, which correlate within a year to the timing of peaks in the annual plant component of CO_2.

Figure 11.4 Annual changes in atmospheric CO_2 concentrations with ocean and plant components of CO_2 in the atmosphere, averaged for nine stations

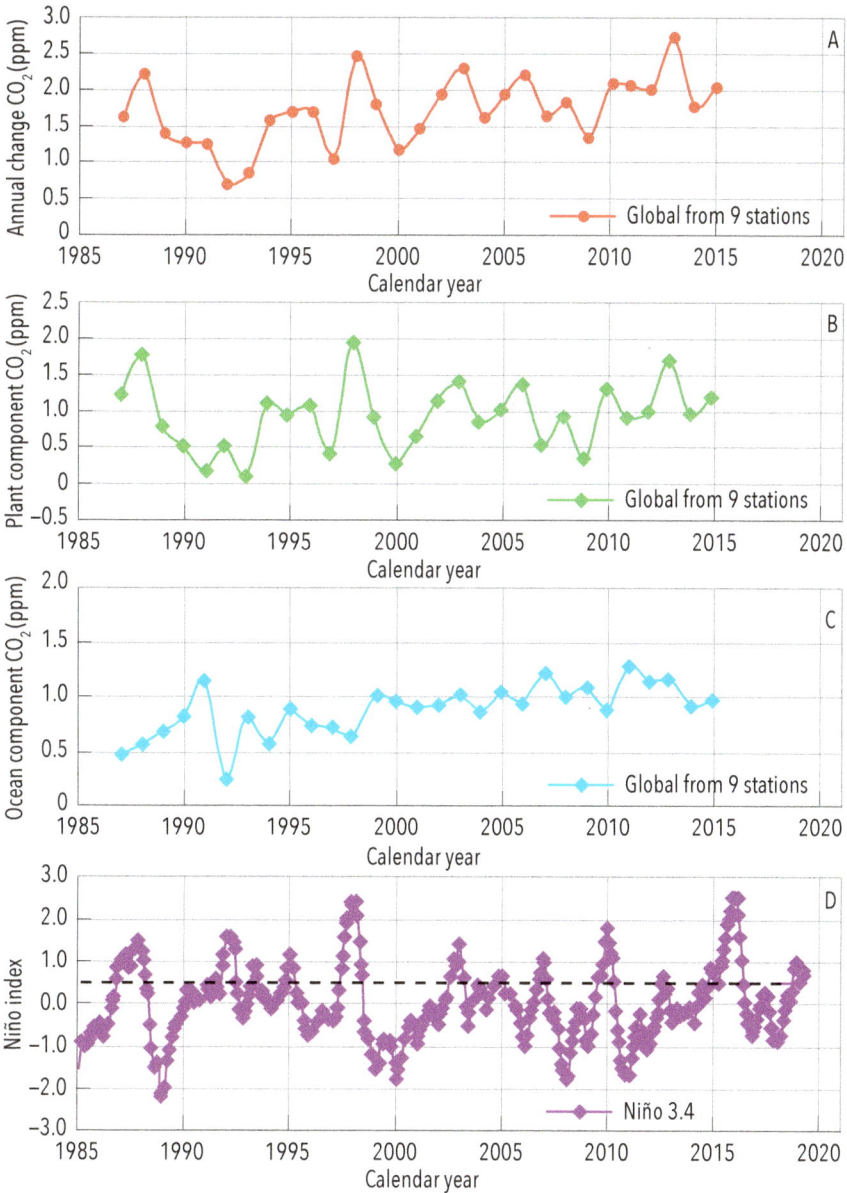

Annual changes in atmospheric CO_2 concentrations with ocean and plant components of CO_2 in the atmosphere, averaged for nine stations listed in table 7 (Quirk & Asten 2023). (a) Atmospheric CO_2 concentrations, (b) plant component CO_2, (c) ocean component CO_2 and (d) Nino 3.4 index in purple with dashed line to show when index is above or below 0.5.

Source: Quirk and Asten 2023.

Conclusion

During the time span studied, 1978 to 2015, the two different atmospheric CO_2 concentrations – having separate origins: one from deep ocean sources and the other from plants, including fossil fuel emissions – show a differing geographical distribution of the components. This indicates that variations in deep ocean and plant components of atmospheric CO_2 concentrations are attributable to independent mechanisms.

The ocean component and plant components, as derived from isotopic data, are approximately equal at 50% of the annual CO_2 concentration increase. This result is similar to the outcome of a recent study by Holzer and DeVries (2022) using a very different methodology – an ocean circulation model (without isotopic data) to track emitted carbon through the atmosphere–ocean system from pre-industrial times to the present (1780 to 2020). That study provided an estimate of 55% ocean source in atmospheric CO_2. Our approach and that of Holzer and DeVries yield estimates that are similar within a few percentage points.

Four essential conclusions may be drawn from this study:

1. Annual increases in atmospheric CO_2 concentrations from 1978 to 2015, when split into plant and deep ocean sources, show approximately equal parts when averaged over all latitudes.

2. The deep ocean component of CO_2 has its maximum value at the Equator. We conclude that it is an independent source of atmospheric CO_2. This is further demonstrated by its limited year-on-year variability. We hypothesise that the observations of changes in ocean-source CO_2 are attributable to the upwelling of CO_2-rich water transported across the sea floor via the global thermohaline circulation of the oceans.

3. The plant-source CO_2 in the atmosphere shows strong seasonal variability in northern latitudes up to the Arctic Circle and more temporal year-by-year variability than that of the ocean-source CO_2, and shows maximum variations coincident with the timing of El Niño events. It also shows a change in trend associated with the 1995 AMO phase change. These spatial and temporal variations are consistent with

the plant-source CO$_2$ being associated with multiple variations in fossil fuel CO$_2$ emissions, together with natural variations in the net primary productivity of oceanic phytoplankton, and forest and peat fires.

4. The differing variations by year and by latitude of plant-source and deep ocean-source CO$_2$ suggest that inclusion of such isotopic data in full carbon cycle models for the Earth may provide additional constraints on the modeling of bidirectional atmosphere–ocean transfers of CO$_2$.

Acknowledgements

This is an abridged version of the original paper by Quirk and Asten (2023).

12 Warming by Carbon Dioxide is Too Small to Matter, According to Will Happer

Dr Jennifer Marohasy

Not many scientists have the expertise to work through the mathematics to verify the claim that increasing atmospheric concentrations of carbon dioxide (CO_2) will cause catastrophic global warming. However, William Happer, an expert in radiation transfer and emeritus professor of physics at Princeton university, can. His conclusion is that the basic physics of global warming theory, and the radiation transfer calculations as done by the Intergovernmental Panel on Climate (IPCC), are correct; they indicate that even a doubling beyond current concentrations will only cause about a 0.71 °C increase in temperatures – a figure that is too small to matter. What most of us don't realise is that all of the 'catastrophe' in the IPCC models derives not from radiation transfer, but from assumed huge positive feedbacks that Happer considers are 'rather puffed-up'.

In September 2023, Happer gave a series of lectures in Australia at the invitation of the Institute of Public Affairs. This chapter explores the issues he raised, including how increasing concentrations of CO_2 are 'greening' the Earth. Of most interest, though, given his expertise, are Happer's conclusions on CO_2 as a greenhouse gas, and on global warming theory being central to claims that unless we reduce emissions of CO_2, we risk a climate catastrophe.

Catastrophes aside, the Earth is greening – so why don't we care?

Equilibrium climate sensitivity – how much temperatures will eventually increase if there is a doubling of atmospheric levels of CO_2 above pre-industrial levels – sits at the heart of the climate models, which drive the public policies that demand economic reform. Perhaps because Happer has worked through the mathematics and concluded that even a doubling of CO_2 cannot cause a climate catastrophe, he is keen to discuss its benefits. Having slain the 'boogeyman', he has moved on to what he sees as an undeniable benefit for plant growth caused by increasing atmospheric concentrations of CO_2 – the greening of the Earth.

Happer concluded each of his lectures by explaining how CO_2 drives life on Earth. The growth of plants depends on CO_2, consequently, the fact that atmospheric concentrations are increasing is a good thing.

As he explained, all food chains begin with plants. Carbon dioxide, which is a natural trace gas in the air, diffuses into the leaves of plants through little holes called stomata. Within the leaves, specifically within the chloroplast, the CO_2 combines with water, and – as long as there is energy provided by the Sun and enough water – it is then converted into simple carbohydrates: one of the three main nutrients found in foods.

Some four billion years ago, CO_2 made up nearly 40% of the total atmospheric mass of the atmosphere. Considering the last half billion years, the global atmospheric CO_2 concentration has shown a general decreasing trend; at present that concentration is very low. So while atmospheric concentrations have increased by about 35% over the last 200 years, this is from a very low base. Carbon dioxide at the present time is much lower in concentration than has prevailed over most of geological history, when levels have been two or three times greater than they are now. The declining trend over time is shown in Figure 4.4 on page 78. This is not Happer's area of expertise but something he is keen to explain; moreover, the science is not contentious.

Happer also explains that plants photosynthesise most efficiently when CO_2 levels are higher, much higher, which is exactly why CO_2 levels in greenhouses are artificially increased. Plants are sensitive to even small increases in atmospheric levels of CO_2. The increase from perhaps 280 parts per million (ppm) 200 years ago to a little over 400 today has, according to Happer, caused 'greening' around the world. This claim is supported by the scientific literature (see Idso 2017, for a review).

Happer correctly explained in his lecture series that this greening is being recorded by satellites and being measured as the Normalized Difference Vegetation Index (NDVI). The trend, as measured by NASA, is one of increasing greenness.

Chapter 13, of *Climate Change the Facts 2017* by Craig Idso is entitled 'Carbon dioxide and plant growth' and includes a table showing the mean percentage biomass increase for common crops given a 300 ppm increase in CO_2. From what is essentially a doubling of the amount of CO_2 in the air, the increase in productivity amounts to about one-third, on average, which is by no means trivial.

This good news that Happer focuses on is not included in Greta Thunberg's *The Climate Book*, nor in any other bestsellers. Perhaps this greening of our planet could be accepted as a corollary to increasing concentrations of CO_2 if activists and scientists alike understood why it is that Happer concludes warming by CO_2 is too small to matter – but this requires some understanding of the physics of radiation transfer.

All in agreement, the physics of radiation transfer

Just as the satellites are able to measure the amount of greening on Earth, they are also able to measure how much radiation is emitted by the Earth at different wavelengths. The most important slide in Happer's Australian lectures showed the amount of radiation emitted to space from Earth at different frequencies (see Figure 12.1). Most importantly, it showed that a doubling of CO_2, specifically increasing the concentration from 400 ppm to 800 ppm, will only cause a 1% drop in the amount of thermal radiation lost to space and thus the amount of heat retained, given the increasing concentrations of this greenhouse gas.

The changes in absorption and emissions from doubling CO_2 only decreases radiation out to space by a little more than 1%, as shown in Figure 12.1. Professor Happer explained:

'Doubling CO_2 would decrease thermal radiation to space by about 1%, or about 3 W/m² out of about 300 W/m² already being lost to space. The IPCC gets essentially the same answer.'

Without feedbacks, this radiation decrease can be restored by an increase in absolute temperature of about 0.25%. The absolute temperature scale is measured in degrees Kelvin.

Figure 12.1 Thermal radiation to space from the Earth

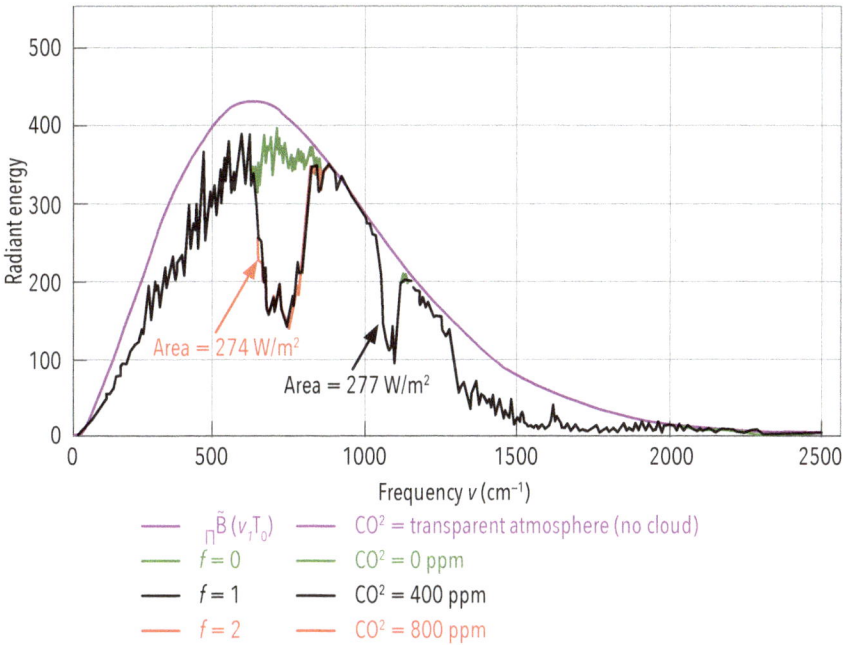

Thermal radiation to space from the Earth, with a surface temperature of 15.5 °C and with greenhouse gases is the area under the jagged black 'Schwarzschild' curve. This is only about 70% of what it would be without greenhouse gases, the area under the smooth purple 'Planck' curve. The Sun heats the Earth and greenhouse gases hinder the cooling.

Source: van Wijngaarden, WA & Happer, W 2020, Dependence of Earth's Thermal Radiation on Five Most Abundant Greenhouse Gases, Fig. 4, viewed 23 October 2023, https://arxiv.org/pdf/2006.03098.

The important point that Happer makes is that most radiation, as shown in Figure 12.1, is from emissions by high-altitude greenhouse gases that are cooler than the surface. Except for frequencies in the atmospheric window, essentially all the surface radiation has been absorbed with none reaching outer space. For sure, it is important to understand that there is emission as well as absorption. It is also important to stress that four times smaller relative increases in temperature are sufficient to compensate for a relative decrease of flux out to space.

The values quoted by Happer are based on theoretical calculations beginning with the mathematical formula that relates the temperature at the surface of the Earth to the amount of heat that it radiates. The mathematics, developed by the German physicists Josef Stefan and Ludwig Boltzmann towards the end of the nineteenth century, is known as the Stefan–Boltzmann law.

This same formula that Happer references is also the formula that his Princeton colleague S. George Philander begins with in his book, *Is the Temperature Rising: The Uncertain Science of Global Warming* (1998), to explain the physics of CO_2 and global warming.

Philander is a much-lauded, mainstream, climate scientist. Philander's book, which was published more than 20 years ago, explains the science of radiative transfer and greenhouse gases in much the same way

Figure 12.2 The Earth's radiation as measured by a satellite over the island of Guam in the tropical Pacific Ocean

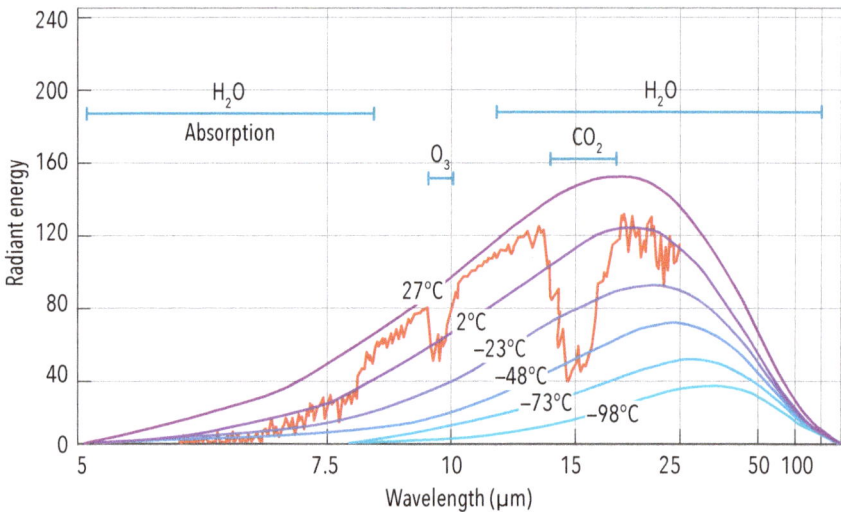

The jagged curve of the Earth's radiation as measured by a satellite over the island Guam in the tropical Pacific Ocean, where the surface temperature is near 300 °K (or 27 °C). The smooth curves show the expected radiation from the surfaces with the indicated temperatures. Much of the measured radiation is from the surface of Guam, except that for certain wavelengths, the radiation appears to be from regions with much lower temperatures, presumably from the upper atmosphere. The atmospheric greenhouse gases that absorb the surface radiation and, in turn, radiate at lower temperatures include: CO_2, at wavelengths centred on 15 microns; water vapour over a broad spectrum of wavelengths; and ozone at wavelengths near 10 microns.

Source: S. George Philander, *Is the Temperature Rising? The Uncertain Science of Global Warming*, 1998, Princeton University Press, p. 51.

that Happer explained them in his Australian lecture series. Philander's book also includes a figure (see Figure 12.2) showing the actual amount of radiation emitted from the Earth as measured by a satellite over the island of Guam in the tropical Pacific Ocean.

The easiest way to distinguish thermally emitted infrared radiation as distinct from sunlight is to observe at night when there is no incoming sunlight. Many geostationary satellites measure Earth's radiation, including visible, near infrared and far infrared thermal wavelengths. The night-time side of the Earth is pitch black for visible and near infrared radiation, but it is just as bright as the daytime side of the Earth for thermal infrared. As Happer describes it, 'The Earth glows in the dark for thermal radiation'.

During the daytime, sunlight heats the Earth and its atmosphere, whereas thermal radiation is emitted to space at nearly the same rate, both night and day. This thermal radiation is heat, emplaced by the daytime sun, which has been transformed back into thermal radiation at longer wavelengths by greenhouse molecules, clouds and the Earth's surface.

As Philander explains it:

> The warmer the surface, the more heat it radiates. Temperatures at the surface of Guam are close to 27 degrees Celsius, and part of the observed curve is indeed close to the smooth one that corresponds to a surface at that temperature. The departure from the smooth curve is most striking in a gaping hole centered on a wavelength near 15 microns. The gas carbon dioxide, which absorbs that particular wavelength may be present in modest amounts – it accounts for 0.035% of the atmosphere – but it clearly is an effective absorber of infrared radiation.

In September 2023, as I write this chapter, the annual average CO_2 concentration as measured at Mauna Loa is expected to be almost, but not quite, 420 ppm this year, which is 0.042% of the atmosphere.

Still agreement, saturation and vertical temperature variation

The atmosphere's heating response to increasing concentrations of CO_2 is *not* at all like a plant's response. The physics of the atmosphere is quite different to the physiology of a plant. As explained earlier, with a doubling of CO_2, plant biomass increases significantly – by about one-third, on average. In contrast, the extra forcing in the atmosphere from a doubling of CO_2 is generally agreed to be minuscule by physicists with expertise in radiation transfer, adding up to about 1% in total, or 3 W/m². This is the radiation transfer component, without amplifying feedbacks.

The physics is complicated by the complex structure of the atmosphere, with temperatures alternately decreasing and increasing to define layers known as the troposphere, stratosphere, mesosphere, and thermosphere. Then there is the particular nature of greenhouse gases' molecules causing them to show a logarithmic decline in their absorption characteristics as they rise and approach saturation. And in the lower atmosphere, in the troposphere, CO_2 was already at near saturation even at preindustrial levels even though, as Happer emphasised in this lecture series, these preindustrial levels are low when considered over geological history.

Even at preindustrial levels of CO_2, most of the available long-wave thermal radiation – the heat emplaced by the daytime sun that has been transformed back into thermal radiation – is absorbed in the lower atmosphere by greenhouse gases.

A relatively plain English explanation of this is given by Jack Barrett in an important article published in *Energy & Environment* (2005). Based on analysis of the infrared spectra of the four main greenhouse gases – water vapour, CO_2, methane and dinitrogen monoxide – as calculated using the High Resolution Transmission Molecular Absorption Database (HITRAN) database, Barrett explains that:

> The absorption values for the pre-industrial atmosphere add up to 86.9%, significantly lower than the combined value of 72.9%. This occurs because there is considerable overlap between the spectral bands of water vapour and those of the other greenhouse gases. If the concentration of carbon dioxide were to be doubled in the absence of the other greenhouse gases the increase in absorption would be 1.5%. In the presence of the other greenhouse gases the same doubling of concentration achieves an increase in absorption of only 0.5%, only one third of its effect if it were the only greenhouse gas present. Whether this overlap effect is properly built into models of the

atmosphere gives rise to some scepticism.

The greenhouse gases absorb 72.9% of the available radiance, leaving 27.1% that is transmitted of which an amount equivalent to 22.5% of the total passes through the window [to outer space] and the other parts of the spectral range transmit only 4.6%. For the doubled carbon dioxide case this small percentage decreases slightly to 4.1%. These small percentage transmissions are reduced by 72.9% and 73.4% respectively by the second layer of 100 m of the atmosphere so that only ~1% in both cases is transmitted to the region higher than 200 m.

The lower atmosphere is already at saturation. At higher altitudes, CO_2 is radiating energy to outer space but at a rate proportional to the fourth power of its absolute temperature, that will vary with altitude making the mathematics somewhat complicated. Nevertheless, a knowledge of absorption bands and Schwarzschild's equation for radiative transfer, allow Barrett, Happer and other physicists to make calculations, and compare these with actual emission spectra of the Earth as measured by the satellites at different latitudes and altitudes.

It is devastating to the central thesis of global warming theory that both the satellites and the mathematics show even a doubling of atmospheric levels of CO_2 is too small to matter; consequently, this information is mostly ignored in the IPCC's version of physics.

Conclusion

Happer's commentary on the effects of CO_2 on Earth's plant life, which he has made repeatedly – not just during his Australian lecture series, but more generally over the last decade – can be measured by satellites and explained in terms of plant physiology. The great deserts of the Earth are shrinking, not growing. And they are shrinking because of the relatively small increase in atmospheric levels of CO_2 over the last 200 years.

It is also clearly stated by Happer, but not at all by his IPCC physicists' colleagues, including other physicists at Princeton and other Ivy League universities, that modern analytical and theoretical physics does *not* support the popular notion that more CO_2 will absorb significantly more heat leading to a meaningful increase in global temperatures. This is what we have been led to believe, with the analogy continually made that adding more CO_2 is like adding another blanket to a bed. The layering of the atmosphere is in fact infinitely more complicated than the making of a bed, and even quite different to the physiology of photosynthesis in plants. All of this does matter, because government policies causing major economic upheaval are actually based on a quite different understanding of the science.

13 Volcanoes and Climate
Dr Arthur Day

The recent spectacular volcanic eruption at Tonga came with very little warning and should remind us that we have no power over the timing or impacts of such events. In this chapter, beginning with an account of what took place at Tonga, I give examples of even bigger eruptions in the past and contemplate just what kind of climate change we should really be worried about – gradual global warming or sudden unexpected global cooling triggered by volcanoes?

When volcanic eruptions eject enough material into the atmosphere, they can potentially affect the climate. It has been directly observed that even some moderate eruptions can trigger rapid cooling. Ice cores containing a record of temperatures over time confirm that the coldest decades of the past 2500 years are associated with volcanic episodes (Sigl et al. 2015). During an eruption, volcanic material injected into the stratosphere,[1] the second layer in the atmosphere, can remain aloft for a very long time, Figure 13.1. Within a few weeks, a thin wispy cloud of very fine dust and gasses can gradually spread around the entire globe,

---

[1]   The stratosphere, at 15–50 km high, is the second layer in the atmosphere above the troposphere. It is a very stable, cold, calm and dry place. 'Weather' and vertical air movement is almost absent. There is no precipitation, the air is very thin, and the wind blows quickly and strongly without turbulence. An important difference between the stratosphere and the humid and turbulent troposphere beneath is that very fine dust and volcanic gasses can be rapidly flushed from the troposphere due to regular rain, but in the stratosphere, they remain aloft for many months or even years, so they can have lingering climate impacts.

Figure 13.1 Photograph from the Space Shuttle above South America (Mission STS-43) on 8 August 1991

Two months after the eruption of Mount Pinatubo in the Philippines, a thin double-layer cloud of eruption products (dark streaks) remained suspended in the stratosphere above South America. The persistent thin cloud reduced the amount of sunlight reaching the Earth's surface, leading to lower global temperatures for more than 15 months.

Source: Photo courtesy of the Earth Science and Remote Sensing Unit, NASA Johnson Space Center, photo ID STS043-22-12.

reducing the amount of sunlight reaching the ground worldwide, leading to global cooling.

'Big' eruptions witnessed in recent decades are only relatively small in comparison to those that geology tells us have occurred in the past and will happen again. Cooling impacts of very big eruptions have been sufficient to cause continent-wide crop failures and widespread famines. Even farther back in time, super-eruptions may have triggered mass extinctions that altered the course of evolution.

The extent of the cooling, and how long it persists, depends on the amount of material ejected (the size of the eruption), what the plume is made of, how high it gets, how long it stays up there, and the latitude of the eruption. Even modest eruptions can eject billions of tonnes (and cubic kilometres) of volcanic material tens of kilometres into the sky.

The contents of the volcanic cloud can vary widely according to the circumstances of the eruption, but it will consist mostly of explosively pulverised rock ('volcanic ash'), water (H_2O) vapour, carbon dioxide (CO_2), and lesser amounts of sulphur dioxide (SO_2) and hydrogen chloride (HCl). While H_2O and CO_2 from volcanoes are greenhouse gasses (GHG) that could add to warming, direct observations repeatedly show that net cooling takes place instead. This is because very fine ash particles, and the gasses SO_2 and HCl, have a much greater cooling impact. They can act as powerful agents that alter the passage of heat through the atmosphere. The two gasses referred to are highly chemically reactive and the chemical changes they undergo can magnify their effect. For example, SO_2 rapidly combines with water vapour to turn into fine sulphuric acid droplets, which then become suspended sulphate particles. Sulphate particles are the main cause of cooling from volcanic eruptions. They strongly reflect the sun's rays back into space, but they also scatter and absorb some as well. The absorption heats the stratosphere layer but the combination of reflection, scattering, and absorption also results in less sunlight getting through, which leads to a cooling effect below.

HCl works differently to SO_2. Chemical reactions involving HCl and water vapour can create chlorine (Cl) compounds that destroy the ozone (O_3) molecules in the ozone layer. Ozone is a GHG and weakening the ozone layer will allow more heat to escape the upper atmosphere, also contributing to cooling. Thus, the reactive effects of these gases can have significant knock-on climate-cooling impacts.

Hunga Tonga–Hunga Ha'apai January 2022 – no measurable global temperature impact

On 15 January 2022 at 5:14 pm the final phase of an eruption began that culminated in the largest volcanic explosion this century, and one of the biggest in recorded history. Hunga Tonga–Hunga Ha'apai is a violent undersea volcano 20 km in diameter across at its base and rising 2000 m above the sea floor. It sits along the shallow Tonga–Kermadec Ridge and is a stratovolcano, a type of violent volcano that grows above unstable zones where the Earth's crust is being destroyed.

Before the eruption, the volcano's 4×2 km caldera was only about 150 m beneath the surface of the sea. It is this shallow seawater that makes

Figure 13.2 Himawari-8 satellite image of the 15 January 2022 eruption of Hunga Tonga-Hunga Ha'apai, taken an hour and a half after the eruption began

Source: Japan Meteorological Agency, CC BY 4.0 https://creativecommons.org/licenses/by/4.0, via Wikimedia Commons. Animated GIF URL: https://upload.wikimedia.org/wikipedia/commons/c/c6/ Tonga_Volcano_Eruption_2022-01-15_0320Z_to_0610Z_Himawari-8_visible.gif.

eruptions from volcanoes of this type so powerful. Because of the lack of pressure, shallow seawater can be vaporised by an eruption into the atmosphere far more readily than in deeper seawater. The latest phase of activity commenced with a relatively modest eruption on 20 December 2021. As this activity subsided, the volcano was declared dormant again on 11 January. But, in a stark illustration of the unpredictability of volcanoes, it erupted again only a few days later on 15 January with a second explosion about seven times more powerful than the first. This huge explosion was likely caused by seawater entering a shallow chamber filled with very hot magma located just beneath the submerged caldera. An additional factor that makes this volcano more explosive than most others in the Pacific is that its magma has more silica, making it more viscous and therefore more prone to explode when sudden depressurisation releases the dissolved gasses.

Figure 13.3 Eruption of the Hunga Tonga–Hunga Ha'apai underwater volcano on 15 January 2022

Photo credit: Tonga Geological Services.

The main 15 January eruption lasted twelve hours and released 7–10 km³ of ejecta into the atmosphere. The initial atmospheric plume rapidly rose from the ocean surface and within about 30 minutes it had ascended to a height of 58 km, passing through the stratosphere to penetrate about 7 km into the base of the mesosphere, the third layer of the atmosphere.[2] NASA reports this was the highest eruption plume ever observed in the satellite record.[3]

In less than two hours, a giant umbrella-shaped cloud had spread to a width of 500 km at its widest point about 30 km above the sea, well into the stratosphere. Once in the stratosphere the plume materials remained in suspension and encircled the globe within two weeks. As a result, after Tonga, the suspended very fine ash led to spectacular orange sunsets accompanied by deep purple dawn and dusk skies around the world for many months.

---

2   The mesosphere is the third layer in the atmosphere 50–100 km high. Lying above the stratosphere, it extends to near the edge of space.

3   https://earthobservatory.nasa.gov/images/149474/tonga-volcano-plume-reached-the-mesosphere *NASA*. 17 February 2022.

Despite its violence, the Tongan eruption did not have any discernible impact on the climate. This might be linked to the plume being very low in SO_2 – the main volcanic emission that causes cooling. In a study published in *Nature Climate Change*, Jenkins et al. (2023) estimated that Tonga only injected 420,000 tonnes of SO_2 into the stratosphere. This might sound a lot, but it was only about 2% of the amount injected during Mount Pinatubo's eruption in 1991, which did result in detectable cooling.

An exceptional feature of the Tongan eruption was that it injected 146 million tonnes of water vapour into the stratosphere (Jenkins et al. 2023), more than has ever been observed in the satellite era. This might explain the eruptive plume's impressive height because the high water-vapour content would have made it less dense than it otherwise would have been, making it more buoyant. Jenkins et al. (2023) estimated that the eruption increased the water-vapour content of the global strato-sphere by around 10%, and it might remain there for at least a year. Theoretically, because water vapour is a GHG, this could trigger a cycle of cooling in the stratosphere and warming in the troposphere, but no clear global warming signal has been measured. However, using a model, Jenkins et al. (2023) calculated the extra water vapour could increase the global temperature by up to 0.035 °C. This is a meaninglessly small number, even if it were an actual measurement. Any weak warming effect of the water vapour might be cancelled by the cooling effects of the SO_2, making any climate impacts of the Tongan eruption too small to detect.

The eruption delivered another surprise. A week afterwards, O_3 above the tropical southwestern Pacific and Indian Oceans had decreased by 5%. Evan et al. (2023), in a study published in *Science*, reported they were able to measure this change directly by launching a balloon into the stratosphere above Réunion Island in the Indian Ocean, more than 12,000 km to the west of Tonga. The reduction was caused by O_3-depleting chemical processes fed by HCl in the presence of the large amount of water vapour from the (salty) seawater that was injected into the stratosphere.[4] This was a very important finding because O_3 protects

---

4   Chemical reactions between HCl and H_2O in the eruption cloud create chemically active Cl monoxide (ClO). This compound acts as a 'catalyst', an accelerant for the breakdown of O_3 without undergoing any permanent chemical changes itself. In this way, the original HCl's influence is massively amplified because each new ClO molecule is able to break down many thousands of O_3 molecules.

us from harmful ultraviolet (UV) rays as well as being a valuable GHG in the stratosphere.

Mount Pinatubo June 1991 – a measurable temperature impact

Prior to the Tongan eruption, the largest volcanic plume in the satellite era came from Mount Pinatubo in the Philippines. After being dormant for 500 years this volcano ejected 10 km³ of ash and gasses into the stratosphere to an altitude of 35 km. The main eruption lasted just nine hours.

The Pinatubo plume was only about two thirds the height of the Tongan one but within a few hours it had injected 14–22 million tonnes of SO_2 into the stratosphere (Guo et al. 2004), whereas the Tongan eruption only injected about a 50th as much. The relatively high amount of fine suspended sulphate particles related to Pinatubo led to strong reflection, absorption, and scattering of some of the sunlight. This resulted in a sudden reduction in direct transmission of solar radiation through the atmosphere (Figure 13.5a) while the absorption also led to

Figure 13.4 The climate-changing Mount Pinatubo volcano erupting

Photo credit: David Harlow, US Geological Survey.

Figure 13.5 Effects over time of some recent volcanic eruptions on transmission of sunlight through the atmosphere, and their impact on global temperatures

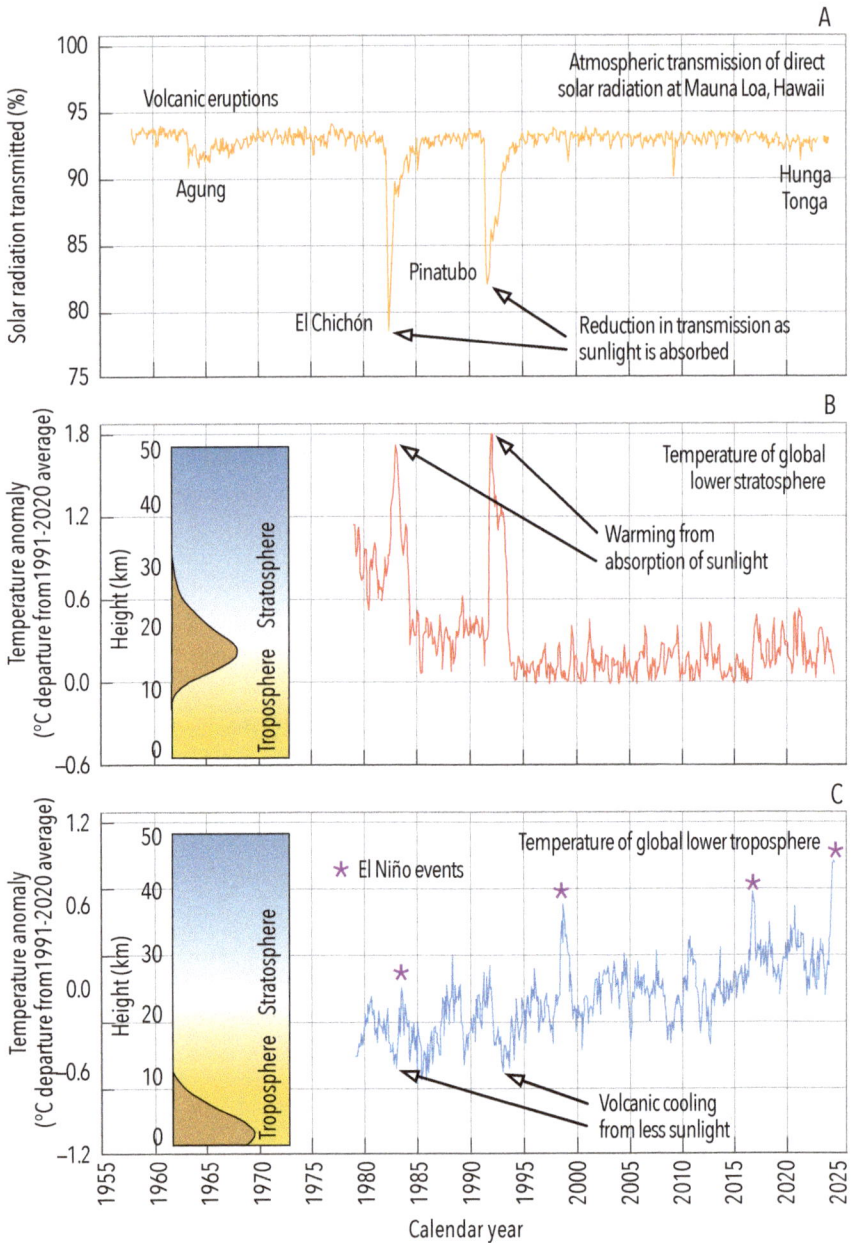

an *increase* in the temperature of the lower stratosphere by as much as 1.5 °C, as shown in Figure 13.5b. The combination of reflection, absorption, and scattering from the suspended sulphate resulted in less sunlight reaching the troposphere and the Earth's surface below, leading to *cooling* (Figure 13.5c). According to NASA, global temperatures cooled by as much as 0.6 °C over the next 15 months.

The cloud from the eruption persisted in the stratosphere for three years, during which the mid-latitude O_3 concentrations fell by 25% to the lowest levels ever recorded (Hofmann et al. 1994).

El Chichón March 1982 – an apparent impact;
Mount Agung March 1963 – no apparent impact

El Chichón came alive after 600 years of dormancy, causing the worst volcanic disaster in Mexico's history beginning on 28 March and continuing until 4 April 1982. While the total volume of material ejected by the eruption was relatively small at less than 1 km³, a tenth of Pinatubo, the plume still reached 27 km and went around the world in three weeks. It injected 12 million tonnes of SO_2 into the stratosphere (McCormick & Swissler 1983), a relatively large amount for an eruption of only modest proportions. This was because its magma was unusually rich in sulphur. Figure 13.5 shows El Chichón had almost the same effects on atmospheric transmission and global temperatures as Pinatubo did, cooling the Northern Hemisphere by about 0.4 °C. The eruption also damaged the ozone layer.

Previous page: The three panels represent time series trends of (A) ground-based monthly averaged measurements of the transmission of solar radiation from 1958 to 2023 at Mauna Loa, and (B) and (C), trends in monthly averaged global temperatures of the lower stratosphere (B) and lower troposphere (C) from 1978 to 2023. The temperature calculations have been adjusted to average zero for the 1991–2020 period to make it easier to visualise changes ('anomalies') in temperature over time. The small graphics at left illustrate the altitude distribution of the satellite measurements used to derive temperatures, depicted as areas in brown under the curves.

Sources: For top panel, the data source was the Global Monitoring Laboratory run by the US National Oceanic and Atmospheric Administration (NOAA). It can be downloaded at https://gml.noaa.gov/grad/mloapt.html. For the middle and bottom panels, the monthly averaged global temperatures in the plots are Version 6 of the University of Alabama in Huntsville (UAH) global satellite temperature dataset. The data files are provided by Dr Roy Spencer and can be downloaded from his monthly temperature commentary page at https://www.drroyspencer.com/latest-global-temperatures/.

On 17 March 1963, the Mount Agung volcano on Bali erupted sending debris 8 to 10 km into the air (possibly not quite reaching the stratosphere). There were no satellite temperature records at this time, and it only produced a very small dip in atmospheric transmission, as shown in Figure 13.5a. As for Hunga Tonga, it is unlikely there was any global temperature impact from this eruption.

Mount Tambora April 1815 – triggered mass starvation

The largest volcanic eruption ever recorded in modern times was the 10 April 1815 eruption of the formerly dormant Mount Tambora volcano on Sumbawa Island in Indonesia. The eruption expelled as much as 150 km³ of magma with a plume height of 40 km. It was at least ten times bigger than Mount Pinatubo. It may have exceeded the size of any other known eruptions in the last 10,000 years. An estimated mass of about 60 million tonnes of SO_2 was ejected (Raible et al. 2016). After the eruption, a caldera measuring 6–7 km across and 600–700 m deep was left, with the height of the volcano reduced by 1250 m.

1816 was 'the year without a summer' as temperatures dropped. Tambora had one of the highest death tolls of any volcanic eruption. It killed about 11,000 people immediately, but the resulting volcanic winter killed more than ten times as many as that, all around the world, due to starvation caused by multiple crop failures. Close to the epicentre, in what was left of Sumbawa Island, Lombok and Bali, people died of hunger and disease as agriculture was destroyed by ash deposits and lack of sunlight. Elsewhere, there were famines in Europe, western China, and North America as crop failures and food shortages took place. In New England there were frosts and snowfalls in June and July and there was a westward migration of New England farmers after their crops failed.

Mount Toba – a prehistoric super-eruption

History has never recorded a supervolcano eruption but such volcanoes are known in the geological record. One example is Mount Toba in Sumatra, which erupted 69,000–77,000 years ago and may have been the most powerful eruption in the last two million years. Today, Toba's caldera is 100×30 km and half a kilometre deep. The eruption that created it is estimated to have only lasted 9–14 days but could have been 100 times more powerful than Tambora, or 1000 times more powerful than Mount Pinatubo or the recent Hunga Tonga–Hunga Ha'apai eruptions. Ice-core evidence suggests average air temperatures worldwide plunged 3–5 °C after the eruption, causing a global volcanic winter of 6–10 years and possibly a 1000-year-long cooling episode. Some believe that Toba almost extinguished humans.

Prehistoric volcanism that changed the course of evolution

Widespread overlapping eruptions from supervolcanoes in Earth's past may have triggered some of the five known mass extinctions in the geological record. In the End-Triassic Mass Extinction, 201.5 million years ago, half of the Earth's land and sea species suddenly became extinct at a time that coincided with massive clustered volcanic eruptions that extended through what is now South and North America, Africa and southern Europe when these regions were one supercontinent. Precise dating of the volcanic rocks shows they erupted 201.57 million years ago. The fossil record of the mass extinction recorded in closely associated sedimentary rocks coincides exactly with this volcanism.

The Triassic event made way for the rise of the dinosaurs, so volcanism may have changed the course of evolution.

The easily overlooked threat of surprise volcanic eruptions

Volcanic deposits and their dates recorded in ice cores show that significant volcanic eruptions have occurred on average twice a century throughout the last 2500 years and there is nothing in our understanding of geology, volcanology, or climatology, to suggest that significant climate-changing eruptions like those of the past can't happen again. Some of the deadliest eruptions in the last two centuries have come from volcanoes with no historical record of eruption, so forecasting volcanoes that have been dormant for hundreds or thousands of years is especially challenging.

Typically, three out of every four centuries have experienced at least one eruption comparable to Mount Pinatubo. Individual large eruptions are not taken into account in climate projections because of their unpredictability, but the cooling effect of even a modest eruption could reverse a century of global warming.

Many people are concerned the (slight) global warming observed over the past 100 years is unnatural and caused by human activity. But sudden unexpected global cooling triggered by volcanism would be far more difficult to adapt to than gradual mild warming. This risk is being ignored in public discussions because of an obsession with CO_2 when, in fact, modern volcanology tells us that it is the potential climate impact of the volcanic chain just to the north of Australia we should be concerned about.

14 Simulation of Winter Warming after Volcanic Eruptions

Dr Petr Chylek, Dr James D. Klett, Dr Glen Lesins, and Dr Manvendra K. Dubey

Major volcanic eruptions produce an aerosol cloud that generally lowers the mean global temperature. However, after major eruptions, winter warming in the latitudinal belt 50 to 70°N has been observed. Such warming was especially high over Eurasia for two winters following the large volcanic eruptions of El Chichon in 1982 and Mt Pinatubo in 1991.

The ensemble means of current state-of-the art climate simulation models that are part of CMIP6 (Coupled Models Inter-comparison Project phase 6) reproduce the winter warmings after Mt Pinatubo, however, they do not reproduce the warming after the El Chichon eruption.

Most of the individual CMIP6 models do reproduce the two winter warmings after Mt Pinatubo, but only a few models can reproduce the observed winter warming after El Chichon. More than half of the individual climate models indicate a cooling where warming is actually observed after the El Chichon volcanic eruption.

This suggests caution in using any individual climate model, or an ensemble of all CIMP6 models, for the study of specific climate phenomena, not only the effect of future volcanic eruptions, but also large forest fires, nuclear war, or possible aerosol seeding for climate modification. Instead, it would be preferable to use an average of a few models that have been tested for the variable of interest and found to have produced simulations of relevant past events that are in agreement with the observations.

Introduction

Large volcanic eruptions emit volcanic gases, mostly water vapour, carbon dioxide (CO_2), and sulphur dioxide (SO_2), that reach the lower part of the stratosphere. SO_2 is oxidised within a few days to form sulphate aerosols (Robock & Mao 1992; Robock 2000). A localised volcanic aerosol layer in the lower stratosphere is spread over much of the globe by stratospheric circulation. This aerosol may lower the global tropospheric and near-surface temperature by the partial scattering of solar radiation back to space. However, an aerosol layer in the stratosphere also induces lower stratospheric heating due to the absorption of solar near-infrared and terrestrial longwave radiation.

Large volcanic eruptions have contributed to past climate variability of the Northern Hemisphere, and have sometimes caused major famines and pandemics, as well as economic disruptions, monsoon enhancement, and general North Atlantic climate variability. Large volcanic eruptions have also caused a wintertime surface warming over the northern part of the Northern Hemisphere, especially over Eurasia. Such warming has been reproduced by several modelling studies (Bittner et al. 2016; Zambri & Robock 2016). However, other researchers using climate models reported no post-eruption winter warming (Stenchikov et al. 2006; Driscoll et al. 2012; Charlton-Perez et al. 2013).

Stenchikov et al. (2006) and Driscoll et al. (2012) analysed the response to nine large volcanic eruptions within the period 1850–2005 during the two winters subsequent to the eruption. They averaged together the first and second winter responses and concluded that models mostly failed to reproduce the observed response. They had hoped to find at least one CMIP5 (Coupled Models Intercomparison Project phase 5) model that would be consistent with observations but were disappointed (Driscoll et al. 2012).

Subsequently, Zambri and Robock (2016) pointed out that averaging over nine eruptions when some eruptions were considerably weaker than others would weaken the overall volcanic signal. Therefore, Zambri and Robock considered only the two strongest eruptions (Krakatau and Mt Pinatubo) and analysed only the first winter immediately after the eruption. By considering only the first winter warming and only the two strongest volcanic eruptions they concluded that climate models in the CMIP5 ensemble were able to produce temperature responses that were in fair agreement with the observations.

In the process of upgrading from the CMIP5 to the CMIP6 class of models, many changes and improvements have been made. In this report, we pose the question as to whether the ensemble of the CMIP6 models can now properly reproduce the winter warming effect of large volcanic eruptions. Here, we consider winter warming during the first and second winters after the eruption by the two strongest eruptions within the second part of the twentieth century, namely El Chichon and Mt Pinatubo. We show that most individual models, as well as the ensemble mean of all CMIP6 models, do reproduce fairly well the temperature variability of northern Eurasia during the two winters after the eruption of Mt Pinatubo. However, the ensemble mean of CMIP6 models, as well as the majority of individual models, produce a significant winter cooling instead of the warming of Eurasia after the El Chichon eruption.

Data and methods

We selected the Mt Pinatubo and El Chichon eruptions because both show strong Eurasian warming in the first and second winters after the eruption. Both eruptions occurred in the northern part of the tropics, and they are both classified as major eruptions. Fortunately, the estimated amount of emitted sulphur is known, and the near-surface temperature data during the time of the eruptions was measured. Both eruptions occurred within the time period of available satellite temperature retrieval of the stratospheric temperature.

We consider the observed surface warming to be likely of volcanic origin because of the observed simultaneous warming of the lower stratosphere due to the absorption of near-infrared solar radiation and longwave terrestrial radiation by the volcanic aerosol cloud. The satellite-retrieved lower tropospheric temperature data are available at the University of Alabama at Huntsville website.[1]

---

1   https://www.nsstc.uah.edu/data/msu/v6.0/tls/uahncdc_ls_6.0.txt

To compare the observed and CMIP6 model-simulated temperature anomaly after the El Chichon and Mt Pinatubo eruptions, we use CMIP6 models listed by the KNMI Climate Explorer website[2] that have at least three ensemble members per model. This leaves us with 29 CMIP6 models. From each model simulation, we use the temperature anomaly with respect to the mean between 1980–1995, which includes both the El Chichon and Mt Pinatubo eruptions. When models provide more than three realisations, only the first three are used in averaging calculations so that all models are treated equally. The model-simulated and observed temperature anomalies are averaged over three months. After that, we calculated the correlation coefficient between the model-produced temperature anomaly and the observed one for a strip over Eurasia (50–70°N and 0–140°E) for a period of 25 months between October 1991 and October 1993 for Mt Pinatubo, and between October 1982 and October 1984 for El Chichon eruptions. These time periods after each eruption contain two winter peaks and the summer in between them. The HadCRUT5 (the latest temperature data assembled by the UK Met Office) is downloaded from the UK Met Office website. The CMIP6 model simulations are downloaded from the KNMI Climate Explorer website.

The observed near-surface temperature anomalies over the area and time period of interest are almost identical among the three most often used temperature records: HadCRUT, NASA GISS (Goddard Institute of Space Studies), and NOAA (Figure 14.1A). All three temperature anomaly records show observed temperature peaks in the two winters following the eruptions. For the following analysis, we use the HadCRUT5.0 data, but any other of the considered temperature sets or their average would lead to identical results.

However, the near-surface temperature data shows not only significant temperature peaks after the eruptions of Mt Pinatubo and El Chichon but also other peaks in between these two eruptions in 1989 and 1990. To determine that other peaks are not also caused by volcanic eruptions, we looked at the temperature profile of the lower stratosphere. Any volcanic aerosol layer reaching the stratosphere is expected to produce a

---

2   https://climexp.knmi.nl/selectfield_cmip6.cgi

Figure 14.1 Temperature anomaly Eurasia and the lower stratospheric and lower tropospheric temperatures

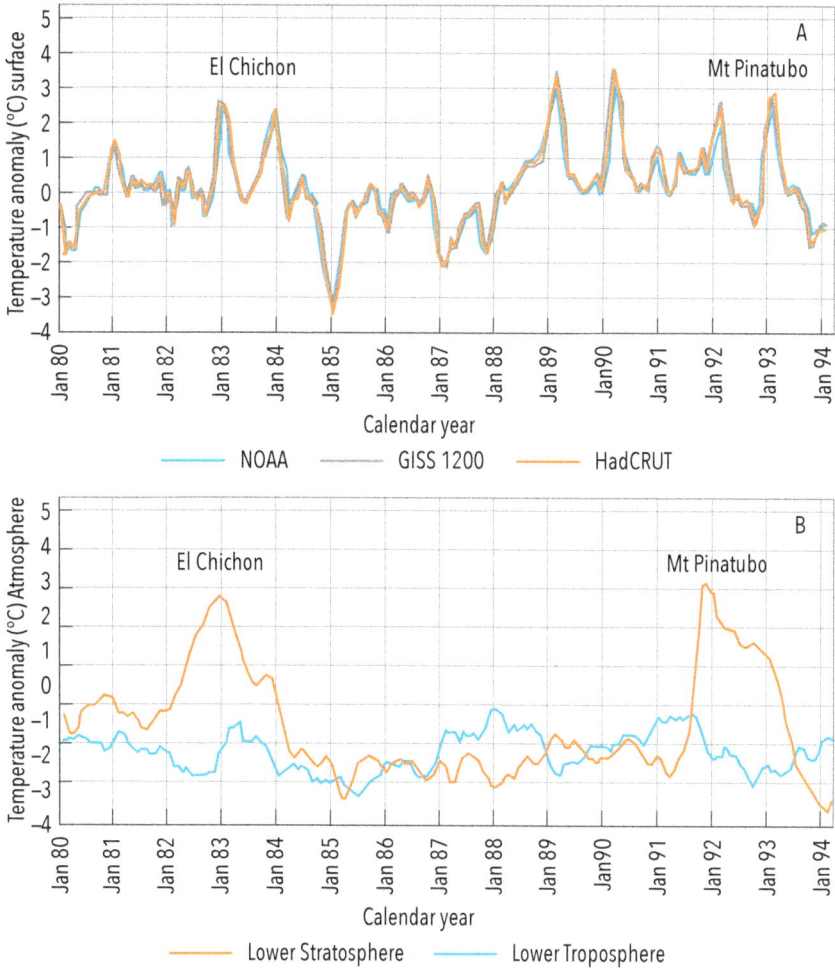

(A) The surface temperature anomalies reported by the HadCRUT, NASA GISS, and NOAA over Eurasia are almost identical. All three datasets show temperature peaks in two winters after the eruptions of El Chichon and Mt Pinatubo. (B) The temperature anomaly of the lower stratosphere (orange curve) and the lower troposphere (blue curve). In the stratosphere, we see statistically significant peaks only after the El Chichon and Mt Pinatubo eruptions.

significant in-situ warming. The satellite data (Figure 14.1B) shows that a lower stratospheric warming is observed within the 1980–1995 interval only after the time of the Mt Pinatubo and El Chichon eruptions. Thus, the two additional peaks between the considered eruptions (Figure 14.1A) are apparently not due to volcanic activity.

Results

The majority of the CMIP6 models do perform well as far as simulating the two winter warmings after the Mt Pinatubo eruption is concerned. Over half (16) of the considered CMIP6 models show positive statistically significant correlations (r>0.40) between the observed and model-simulated temperature anomaly after the Mt Pinatubo volcanic eruption (Figure 14.2A). Another nine models also show positive correlations, which, however, are not statistically significant, and only four models lead to a negative correlation between the observed and model-simulated temperature anomaly after the Mt Pinatubo eruption. The correlation coefficient between the CMIP6 ensemble mean of all models and the observed temperature is statistically significant with the correlation coefficient r=0.62.

However, the situation is different in the case of the El Chichon eruption (Figure 14.2B). Here, most models fail to reproduce the observed warming in the two winters after the eruption. Only three models (ACCESS-ESM1-5, AWI-CM-1-1, and GFDL-ESM4) have a statistically significant positive correlation with the observed temperature anomaly after the El Chichon eruption. Seven of the models show a statistically significant anti-correlation. The correlation coefficient between the CMIP6 ensemble mean of all models and the observed temperature is negative, r=–0.52. More of the CMIP6 models show cooling in the northern part of Eurasia after the El Chichon eruption rather than the observed warming.

To combine the results of the two considered eruptions, we assign a rank order number, from 1 to the model with the highest correlation coefficient, to 29 for the model with the lowest correlation, to each model for both cases of Mt Pinatubo and El Chichon eruptions. By adding these rank order numbers we get the total rank order number for each model (Figure 14.2C).

Figure 14.2 Correlation coefficients for Mt Pinatubo, El Chichon and the CMIP6 models rank order

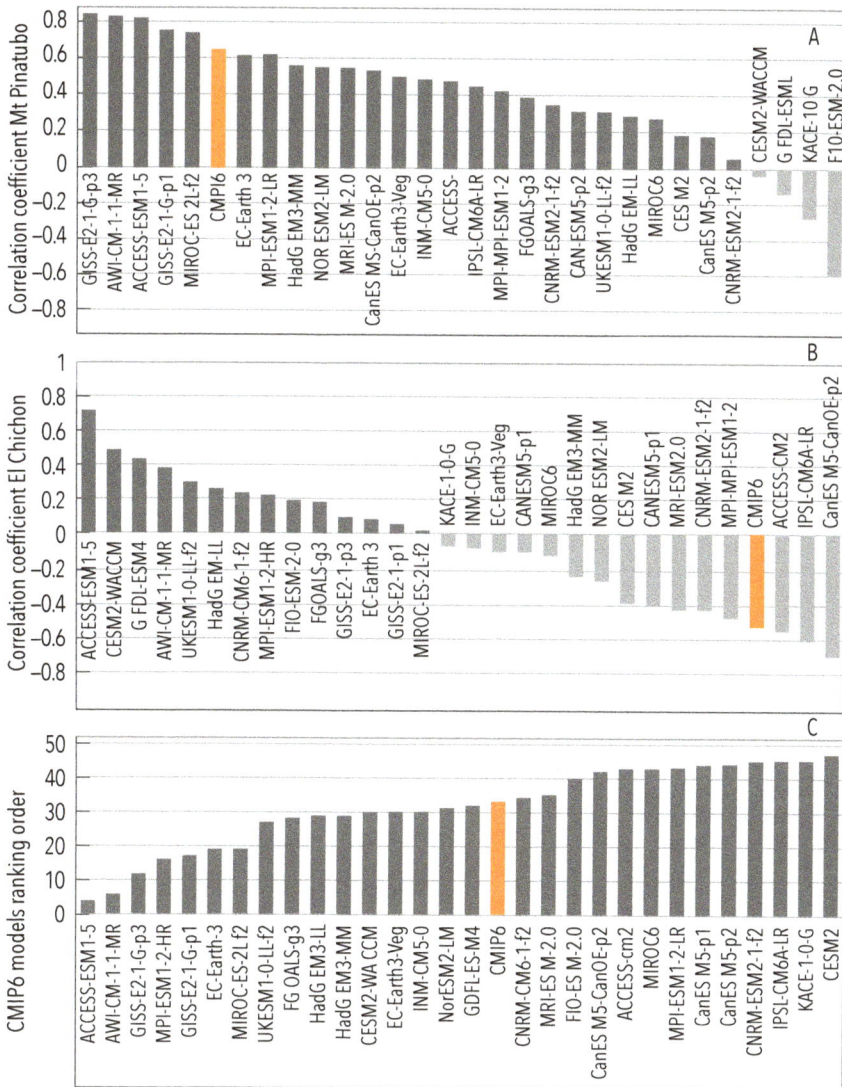

(A) Correlation coefficients between the observed temperature anomaly over Eurasia within 25 months after the Mt Pinatubo eruption. The dark grey columns denote the positive correlations, and the light grey columns the negative ones. The orange column is the correlation coefficient for the ensemble mean of the CMIP6 models. (B) The same after the El Chichon eruption. (C) Rank order coefficient defined in the text for CMIP6 models. The orange column is a rank order coefficient for the ensemble mean of CMIP6 models.

Why some models simulate a cooling instead of the observed warming during the two winters after the El Chichon eruption is not clear, especially since most of the CMIP6 models' responses are in agreement with observations after the Mt Pinatubo eruption. The discrepancy may be at least partially due to the weaker forcing provided by the smaller El Chichon eruption compared to Mt Pinatubo. All the CMIP6 models are supposed to use the same CMIP6-prescribed volcanic aerosol forcing. Consequently, the differences between models' simulations after the volcanic eruptions are due to differences between the ways models process the prescribed forcing.

There are two CMIP6 models – CESM2-WAC-CCM and GFDL-ESM4 – which perform well in the case of the El Chichon eruption (Figure 14.2B); however, they do poorly in simulating the effects of the Mt Pinatubo eruption (Figure 14.2A). This is the opposite of most of the models, which perform well after Mt Pinatubo, but not so well after El Chichon. Why these models do the opposite is not clear.

Two models with significantly lower order numbers than the rest are the Australian model ACCESS-ESM1-5 (Rashid et al. 2022, Zien et al. 2020), and the German model designed by the Alfred Wegener Institute AWI-CM-1-1 (Semmler et al. 2020). The correlation coefficients for the average of these two models and the observed temperature anomaly after the El Chichon and Mt Pinatubo eruptions are 0.64 and 0.89, respectively, compared to the CMIP6 ensemble mean of all models of –0.52 and 0.62.

Ten models with the lowest correlations between the observed and simulated temperature after El Chichon show a positive correlation (winter warming) after Mt Pinatubo, and a negative correlation (winter cooling) after El Chichon. Both correlations are statistically significant (p-value <0.05). Thus, the use of the ensemble means of all the CMIP6 models, or of an arbitrarily chosen individual model, for the assessment of future volcanic eruptions, as well as of possible effects of geoengineering in the form of stratospheric aerosol seeding (Irvine et al. 2019; MacMartin 2016), would not be warranted.

It is also known (Chylek et al. 2022, Hausfather et al. 2022, Chylek et al. 2023a, 2023b) that some of the current climate models (CMIP6), and their ensemble mean, have overestimated mean global warming since about the 1990s. Such imperfect results are of course not surprising, given the fact that the art and science of climate modelling is still very much a work in progress. Some general limitations that remain to be overcome, include: the lack of sufficient resolution to include relevant cloud processes accurately; their use of somewhat arbitrary parameterisations to represent poorly understood but relevant physical processes; the inherent instabilities of the non-linear hydrodynamic equations of motion; and the unavailability of complete initial data. Such deficiencies make it very difficult for current models to simulate the climate with good accuracy over even short periods of time, and the accuracy can be expected to degrade rapidly with the time of projection.

The next question is whether the models' simulations lead to statistically significant peaks. We consider first the temperature anomaly averaged over the best two models with the lowest order number (ACESS-ESM1-5 and AWI-CM-1-1-MR) in Figure 14.3A. The simulations are detrended for the time interval between the two considered eruptions, and the standard deviations are calculated within this interval.

There are only two peaks in the temperature anomaly that are at the 95% confidence level, namely the first peaks after each of the El Chichon and Mt Pinatubo eruptions. No other peaks are at the same confidence level. When the procedure is repeated using four models with the lowest order number (Figure 14.3B), the second peak after the Mt Pinatubo eruption also becomes statistically significant with a confidence level at 95%. For comparison we also show the case for which the ensemble mean of all CMIP6 models is used (Figure 14.3C). It is apparent that the mean of two or four best models, according to the order number, is superior to the ensemble mean of all CMIP6 models.

As noted earlier (e.g. Chen et al. 2020), the amplitude of the winter warming after volcanic eruptions is, in the CMIP6 model simulations, considerably lower than the observed amplitudes. This is, however, at least partially due to the averaging between individual simulations within a model, as well as to smoothing over the three-month period. The amplitude of an individual simulation is on average about three times higher than the amplitude of the average of the four best models.

Figure 14.3 The best two El Chichon models, the best four Mt Pinatubo models and the CMIP6

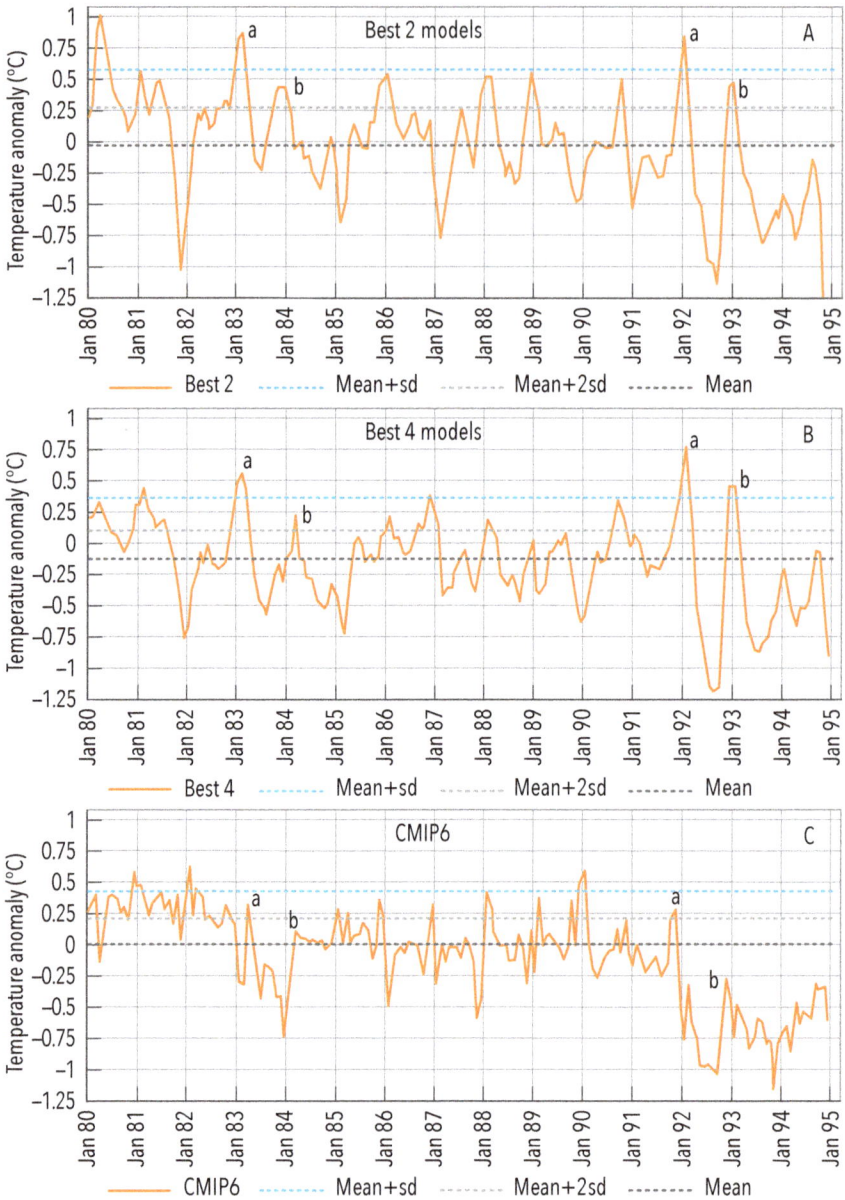

(A) The Eurasia temperature anomaly simulated by the two CMIP6 models with the lowest order number. The data are linearly de-trended in between the two eruptions of El Chichon and Mt Pinatubo (1985–1990). The winter peaks after volcanic eruptions are denoted as (a) and (b). The ---- line is the mean, the ---- line is the mean plus one standard deviation, and the ---- line is the mean plus two standard deviations, all within the six years interval, 1985–1990. (B) Same using the four models with the lowest order number. (C) Same with the ensemble mean of all CMIP6 models.

Thus, to estimate the amplitude of future possible eruptions at the present time, the amplitude of the average of current generation model simulations should be multiplied by about a factor of three.

Discussion and conclusion

Our conclusion is that most of the latest climate models (CMIP6) or their ensemble mean are not suitable for the study of possible effects of future volcanic activities or stratospheric aerosol seedings over northern Europe and Asia. We find that the two or four models (Figures 14.3A and 14.3B) provide much better simulations of the past eruptions of El Chichon and Mt Pinatubo than the rest of the models or their ensemble mean. For any other location, or any other climate variable, the testing of the past simulations has to be compared with observations and the best available model, or a combination of models has to be selected accordingly.

One of the problems with most of the models is that they treat all parts of the zonal belt 50 to 70°N uniformly, while the observed temperature anomaly shows significant differences between Eurasia and North America. The observed data (Figure 14.4A) suggests that there is no correlation (r=−0.10) between mean temperature anomaly over Eurasia and North America. In contrast, the ensemble mean of the CMIP6 simulations (Figure 14.4B) shows a correlation of 0.92.

The CMIP6 simulation seems to be dominated by a seasonal cycle with a superimposed general cooling due to volcanic aerosol, which in the model simulations is similar for both continents. This discrepancy between the observed and simulated temperature is likely due to errors in the models' dynamic responses, as well as to other possible deficiencies, such as a generally poor account given of perturbations in cloud cover and lifetime in response to perturbations in the atmospheric aerosol.

We have to keep in mind that agreement with the past temperature history is, however, no guarantee of correct future projections. Agreement with the past is only a necessary, but not a sufficient, condition for reliable future projection. A model that can reproduce the past temperature correctly may still produce incorrect future projections. So there is even less justification for using a model for future projections that cannot even reproduce past events.

Figure 14.4 The observed temperature anomaly, and the temperature anomaly simulated by the ensemble mean of CMIP6 model simulations over Eurasia and over North America

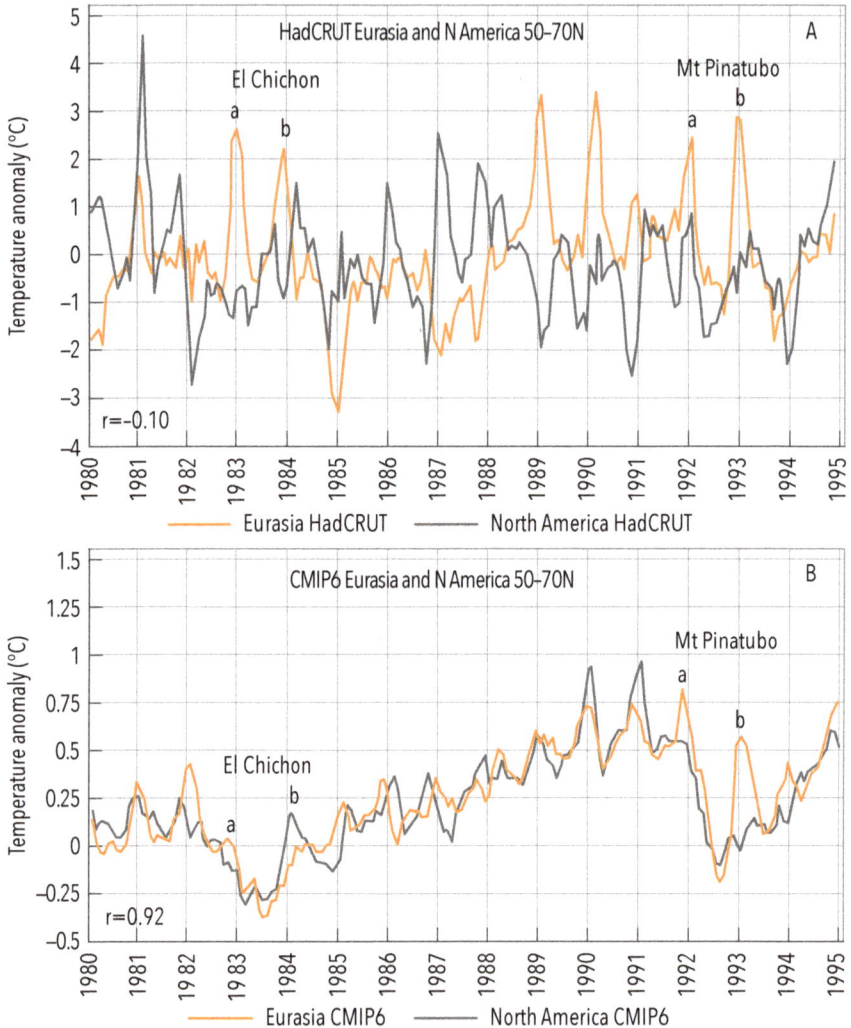

(A) The observed temperature anomaly over Eurasia (orange) and over North America (grey). The winter peaks after volcanic eruptions are denoted as (a) and (b). There is no correlation between them for the two geographical regions (r=−0.10). (B) The temperature anomaly simulated by the ensemble mean of CMIP6 model simulations over Eurasia (orange) and North America (grey). The correlation coefficient between them is r=0.92.

Acknowledgment

The research presented in this report was supported in part by the Laboratory Directed Research and Development program of Los Alamos National Laboratory under project 20200035DR (PI M. Dubey). No new data were used in this work. All the CMIP6 runs are available at the World Meteorological Organization, European Climate Assessment & Dataset, at the link https://climexp.knmi.nl/selectfieldcmip6. cgi?id=someone@somewhere. The observed HadCRUT5 temperature time series were downloaded from https://climexp.knmi.nl/select. cgi?id=someone@somewhere&field=hadcrut5.

References

Foreword

Marcott, SA, Shakun, JD, Clark, PU & Mix, AC 2013, 'A Reconstruction of Regional and Global Temperature for the Past 11,300 Years', *Science*, vol. 339, iss. 6124, pp. 1198–1201.

Introduction

Fanning, AL & Hickel, J 2023, 'Compensation for atmospheric appropriation', *Nature Sustainability* vol. 6, pp 1077–1086, viewed 7 May 2024, https://www.scientificamerican.com/article/rich-nations-owe-192-trillion-for-causing-climate-change-new-analysis-finds/

IPCC 2021, Masson-Delmotte, V, et al. (eds), Climate Change 2021: *The physical science basis. Contribution of Working Group I to the Sixth Assessment Report of the Intergovernmental Panel on Climate Change*, Cambridge, England, Cambridge University Press.

Peel, J 2024, 'Climate litigation is on the rise around the world and Australia is at the head of the pack', University of Melbourne, viewed 7 May 2024, https://findanexpert.unimelb.edu.au/news/63107-climate-litigation-is-on-the-rise-around-the-world-and-australia-is-at-the-head-of-the-pack

1: Cycles in Very Long Temperature Records

Abbot, J 2021, 'Using Oscillatory Processes in Northern Hemisphere Proxy Temperature Records to Forecast Industrial-era Temperatures', *Earth Sciences*, vol. 10, no. 3, pp. 95–117.

Abbot, J & Marohasy, J 2017a, 'The application of machine learning for evaluating anthropogenic versus natural climate change', *GeoResJ*, vol. 14, pp. 36–46.

Allen, MR, Gillett, NP, Kettleborough, JA, Hegerl, G, Schnur, R, Stott, PA, Boer, G, Covey, C, Delworth, TL, Jones, GS, Mitchell, JFB & Barnett, TP 2006, 'Quantifying anthropogenic influence on recent near-surface temperature change', *Surveys in Geophysics*, vol. 27, pp. 491–544.

Auger, JD, Mayewski, PA, Maasch, KA, Schuenemann, KC, Carleton, AM, Birkel, SD & Saros, JE 2019, '2000 years of North Atlantic-Arctic climate', *Quaternary Science Reviews*, vol. 216, pp. 1–17.

Barkhordarian, A, von Storch, H, Zorita, E, Loikith, PC & Mechoso, CR 2018, 'Observed warming over northern South America has an anthropogenic origin', *Climate Dynamics*, vol. 51, pp. 1901–1914.

Bathiany, S, Scheffer, M, van Nes, EH, Williamson, MS & Lenton, TM 2018, 'Abrupt Climate Change in an Oscillating World', *Scientific Reports*, vol. 8, p. 5040.

Black, R, 2008, 'Climate "hockey stick" is revived', BBC News, 1 September, viewed 26 February 2024, website http://news.bbc.co.uk/2/hi/science/nature/7592575.stm

Bond, G, Showers, W, Cheseby, M, Lotti, R, Almasi, P, de Menocal, P, Priore, P, Cullen, H, Hajdas, I & Bonani, G 1997, 'A pervasive millennial-scale cycle in the North Atlantic Holocene and glacial climates', *Science*, vol. 278, pp. 1257–1266.

Bond, GC, Showers, W, Elliot, M, Evans, M, Lotti, R, Hajdas, I, Bonani, G & Johnson, S 1999, 'The North Atlantic's 1–2 kyr climate rhythm: Relation to Heinrich events', Dansgaard/Oeschger cycles and the Little Ice Age, Mechanisms of Global Climate Change at Millennial Time Scales, *Geophysical Monograph Series*, vol. 112, pp. 35–58.

Braun, H, Christl, M, Rahmstorf, S, Ganopolski, A, Mangini, A, Kubatzki, C, Roth, K & Kromer, B 2005, 'Possible solar origin of the 1470-year glacial climate cycle demonstrated in a coupled model', *Nature*, vol. 438, pp. 208–211.

Broecker, WS 2001, 'Paleoclimate: Was the Medieval Warm Period Global?', *Science*, vol. 291, no. 5508, pp. 1497–1499.

Büntgen, U, Tegel, W, Nicolussi, K, McCormick, M, Frank, D, Trouet, V, Kaplan, JO, Herzig, F, Heussner, KU, Wanner, H, Luterbacher, J & Esper, J 2011, '2500 years of European climate variability and human susceptibility', *Science*, vol. 331, pp. 578–582.

Bürger, G, Fast, I & Cubasch, U 2006, 'Climate reconstruction by regression – 32 variations on a theme', *Tellus A*, vol. 58, pp. 227–235.

Christiansen, B & Ljungqvist, FC 2012, 'The extra-tropical Northern Hemisphere temperature in the last two millennia: reconstructions of low-frequency variability', *Climate of the Past*, vol. 8, pp. 765–786.

Christiansen, B & Ljungqvist, FC 2017, 'Challenges and perspectives for large-scale temperature reconstructions of the past two millennia', *Reviews of Geophysics*, vol. 55, pp. 40–96.

Chylek, P, Klett, JD, Dubey, MK & Hengartner, N 2016, 'The role of Atlantic Multi-decadal Oscillation in the global mean temperature variability', *Climate Dynamics*, vol. 47, pp. 3271–3279.

Cook, ER, Buckley, BM, D'Arrigo, RD & Peterson, MJ 2000, 'Warm-season temperatures since 1600 BC reconstructed from Tasmanian tree rings and their relationship to large-scale sea surface temperature anomalies', *Climate Dynamics*, vol. 16, pp. 79–91.

Crowley, TJ & Lowery, T 2000, 'How warm was the Medieval Warm Period?', *Ambio*, vol. 29, pp. 51–54.

Dai, A, Fyfe, JC, Xie, SP & Dai, X 2015, 'Decadal modulation of global surface temperature by internal climate variability', *Nature Climate Change*, vol. 5, pp. 555–559.

De Larminat, P 2016, 'Earth climate identification vs. anthropic global warming attribution', *Annual Reviews in Control*, vol. 42, pp. 114–125.

De Menocal, P, Ortiz, J, Guilderson, T & Sarnthein, M 2000, 'Coherent high- and low-latitude climate variability during the Holocene warm period', *Science*, vol. 288, pp. 2198–2202.

De Vries, H 1958, 'Variation in concentration of radiocarbon with time and location on Earth', *Akademie Van Wet*, vol. 61, pp. 94–102.

Delworth, T, Zeng, F, Vecchi, GA, Yang, X, Zhang, L & Zhang, R 2016, 'The North Atlantic Oscillation as a driver of rapid climate change in the Northern Hemisphere', *Nature Geoscience*, vol. 9, pp. 509–513.

Demezhko, YD & Golovanova, IV 2007, 'Climatic Changes in the Urals over the Past Millennium – an Analysis of Geothermal and Meteorological Data', *Climate of the Past*, vol. 3, pp. 237–242.

Deng, W, Liu, X, Chen, X, Wei, G, Zeng, T, Xie, L & Zhao, JX 2017, 'A comparison of the climates of the Medieval Climate Anomaly, Little Ice Age, and Current Warm Period reconstructed using coral records from the northern South China Sea', *Journal of Geophysical Research Oceans*, vol. 122, pp. 264–275.

Dergachev, VA & Raspopov, OM 2010a, 'Reconstruction of the Earth's Surface Temperature Based on Data of Deep Boreholes, Global Warming in the Last Millennium, and Long Term Solar Cyclicity, Part 2, Experimental Data Analysis', *Geomagnetism and Aeronomy*, vol. 50, no. 3, pp. 393–402.

Dergachev, VA & Raspopov, OM 2010b, 'Reconstruction of the Earth's Surface Temperature Based on Data of Deep Boreholes, Global Warming in the Last Millennium, and Long Term Solar Cyclicity, Part 1, Experimental Data', *Geomagnetism and Aeronomy*, vol. 50, no. 3, pp. 383–392.

Dergachev, VA & Volobuev, DM 2018, 'Solar Radiation Change and Climatic Effects on Decennial–Centennial Scales', *Geomagnetism and Aeronomy*, vol. 58, no. 8, pp. 1042–1049.

Easterbrook, DJ 2016, 'Using Patterns of Recurring Climate Cycles to Predict Future Climate Changes', in *Evidence-Based Climate Science*, 2nd ed.

Eddy, JA 1976, The Maunder minimum, *Science*, vol. 192, pp. 1189–1202.

Egorova, T, Rozanov, E, Arsenovic, P, Peter, T & Schmutz, W 2018, 'Contributions of Natural and Anthropogenic Forcing Agents to the Early 20th Century Warming', *Frontiers in Earth Science*, vol. 6, no. 206.

Esper, FD, Zorita, J & Wilson, R 2010, 'A noodle, hockey stick, and spaghetti plate: A perspective on high-resolution paleoclimatology', *Wiley Interdisciplinary Reviews: Climate Change*, vol. 1, no. 4, pp. 507–516.

Florides, G, Christodoulides, P & Messaritis, V 2010, 'Global Warming: CO2 vs Sun', *Global Warming*, Stuart Arthur Harris (Ed.), InTech, viewed 26 February 2024, http://www.intechopen.com/books/global-warming/global-warming-co2-vs-sun

Folland, CK, Karl, T &Vinnikov, KYA 1990, IPCC First Assessment Report, *Observed Climate Variations and Change*, ch. 7, pp. 195–238.

Folland, CK, Boucher, O, Colman, A & Parker, DE 2018, 'Causes of irregularities in trends of global mean surface temperature since the late 19th century', *Science Advances*, vol. 4, no. 6, eaao5297.

Ge, Q, Hao, Z, Zheng, J & Shao, X 2013, 'Temperature changes over the past 2000 years in China and comparison with the Northern Hemisphere', *Climate of the Past*, vol. 9, pp. 1153–1160.

Georgieva, K, Nagovitsyn, Yu & Kirov, B 2015, 'Reconstruction of the Long Term Variations of the Total Solar Irradiance from Geomagnetic Data', *Geomagnetism and Aeronomy*, vol. 55, no. 8, pp. 1026–1032.

Gervais, F 2016, 'Anthropogenic CO2 warming challenged by 60-year cycle', *Earth-Science Reviews*, vol. 155, pp. 129–135.

Gray, LJ, Beer, J, Geller, M, Geller, M, Haigh, JD, Lockwood, M, Matthes, K, Cubasch, U, Fleitmann, D, Harrison, G, et al. 2010, 'Solar influences on climate', *Reviews of Geophysics*, vol. 48, RG4001.

Haigh, JD 1996, 'The impact of solar variability on climate', *Science*, vol. 272, pp. 981–985.

Haigh, JD 2007, 'The Sun and the Earth's Climate', *Living Reviews in Solar Physics*, vol. 4, pp. 1–63.

Hernández, A, Martin-Puertas, C, Moffa-Sánchez, P, Moreno-Chamarro, E, Ortega, P, et al. 2020, 'Modes of climate variability: Synthesis and review of proxy-based reconstructions through the Holocene', *Earth-Science Reviews*, vol. 209, p. 103286.

Huang, JB, Wang, SW, Luo, Y, Zhao, ZC & Wen, XY 2012, 'Debates on the causes of global warming', *Advances in Climate Change Research*, vol. 3, no. 1, pp. 38–44.

Humlum, O, Solheim, JE & Stordahl, K 2011, 'Identifying natural contributions to late Holocene climate change', *Global and Planetary Change*, vol. 79, pp. 145–156.

Imbers, J, Lopez, A, Huntingford, C & Allen, MR 2013, 'Testing the robustness of the anthropogenic climate change detection statements using different empirical models', *Journal of Geophysical Research: Atmospheres*, vol. 118, pp. 3192–3199.

IPCC 2013, Stocker, TF et al. (eds.), *Climate change 2013: The physical science basis. Contribution of Working Group I to the Fifth Assessment Report of the*

Intergovernmental Panel on Climate Change, Cambridge, England, Cambridge University Press.

Jones, PD 2001, 'Early European Instrumental Records', in PD Jones, AEJ Ogilvie, TD Davies, KR Briffa, KR Kelsey (eds.), *History and Climate: Memories of the Future?* Springer US, Boston, MA, pp. 55–77.

Joos, F & Spahni, R 2008, 'Rates of Change in Natural and Anthropogenic Radiative Forcing over the Past 20,000 Years', *Proceedings of the National Academy of Sciences of the United States of America*, vol. 105, pp. 1425–1430.

Kelsey, AM, Menk, FW & Moss, PT 2015, 'An astronomical correspondence to the 1470-year cycle of abrupt climate change', *Climate of the Past Discussions*, vol. 11, pp. 4895–4915.

Kern, AK, Harzhauser, M, Piller, W, Mandic, O & Soliman, A 2012, 'Strong evidence for the influence of solar cycles on a Late Miocene lake system revealed by biotic and abiotic proxies', *Palaeogeography. Palaeoclimatology, Palaeoecology*, vol. 329–330, pp. 124–136.

Li, L, Wang, B & Zhou, T 2007, 'Contributions of natural and anthropogenic forcings to the summer cooling over eastern China: An AGCM study', *Geophysical Research Letters*, vol. 34, L18807.

Li, C, Zhao, T & Ying, K 2017, 'Quantifying the contributions of anthropogenic and natural forcings to climate changes over arid-semiarid areas during 1946–2005', *Climatic Change*, vol. 144, pp. 505–517.

Liu, Y, Cai, Q, Song, H, An, Z & Linderholm, HW 2011, 'Amplitudes, rates, periodicities and causes of temperature variations in the past 2485 years and future trends over the central eastern Tibetan Plateau', *Chinese Science Bulletin*, vol. 56, pp. 2986–2994.

Ljungqvist, FC 2010, 'A new reconstruction of temperature variability in the extra-tropical Northern Hemisphere during the last two millennia', *Geografiska Annaler Series A Physical Geography*, vol. 92, no. 3, pp. 339–351.

Lüdecke, HJ, Weiss, CO & Hempelmann, A 2015a, 'Paleoclimate forcing by the solar De Vries/Suess cycle', *Climate of the Past Discussions*, vol. 11, pp. 279–305.

Lüdecke, JJ, Weiss, CO, Zhao, X & Feng, X 2015b, 'Centennial Cycles Observed in Temperature Data from Antarctica to Central Europe', *Polarforschung*, vol. 85, no. 2, pp. 179–181.

Lüdecke, HJ & Weiss, CO 2017. 'Harmonic Analysis of Worldwide Temperature Proxies for 2000 Years'. *The Open Atmospheric Science Journal*, 11, 44–53.

Lüdecke, HJ, Cina, R, Dammschneider, HJ & Lüning, S 2020, 'Decadal and multidecadal natural variability in European temperature', *Journal of Atmospheric and Solar-Terrestrial Physics*, vol. 205, p. 105294.

Lüning, S, Schulte, L, Garcés-Pastor, S, Danladi, IB & Gałka, M 2019, 'The Medieval Climate Anomaly in the Mediterranean region', *Paleoceanography and Paleoclimatology*, vol. 34, no. 10, pp. 1625–1649.

Lüning, S, Gałka, M & Vahrenholt, F 2017, 'Warming and cooling: The Medieval Climate Anomaly in Africa and Arabia', *Paleoceanography*, vol. 32, pp. 1219–1235.

Luterbacher, J, Werner, JP, Smerdon, JE, Fernández-Donado, L, González-Rouco, FJ, Barriopedro, D, et al. 2016, 'European summer temperatures since Roman times', *Environmental Research Letters*, vol. 11, p. 024001.

Mann, ME, Bradley, RS & Hughes, MK 1998, 'Global-scale temperature patterns and climate forcing over the past six centuries', *Nature*, vol. 392, no. 6678, pp. 779–787.

Mann, ME, Bradley, RS & Hughes, MK 1999, 'Northern hemisphere temperatures during the past millennium: Inferences, uncertainties, and limitations', *Geophysical Research Letters*, vol. 26, no. 6, vol. 759–762.

Mann, ME, Zhang, ZH, Hughes, MK, Bradley, RS, Miller, SK, Rutherford, S & Ni, FB 2008, 'Proxy-based reconstructions of hemispheric and global surface temperature variations over the past two millennia', *Proceedings of the National Academy of Sciences of the United States of America*, vol. 105, pp. 13252–13257.

Mann, ME, Bradley, RS & Hughes, MK 2009, 'Reply to McIntyre and McKitrick: Proxy-based temperature reconstructions are robust', *Proceedings of the National Academy of Sciences of the United States of America*, vol. 106, no. 6, E11.

Mann, ME, Steinman, BA & Miller, SK 2020, 'Absence of internal multidecadal and interdecadal oscillations in climate model simulations', *Nature Communications*, vol. 11, p. 49.

Margaritelli, G, Cisneros M, Cacho, I, Capotondi, L, Vallefuoco, M, Rettori, R & Lirer, F 2018, 'Climatic variability over the last 3000 years in the central – western Mediterranean Sea (Menorca Basin) detected by planktonic foraminifera and stable isotope records', *Global and Planetary Change*, vol. 169, pp. 179–187.

Masson-Delmotte, V, Schulz, M, Abe-Ouchi, A, Beer, J, Ganopolski, A, González Rouco, JF, et al. 2013, Information from Paleoclimate Archives. In: *Climate Change 2013: The Physical Science Basis. Contribution of Working Group I to the Fifth Assessment Report of the Intergovernmental Panel on Climate Change*, Cambridge University Press, Cambridge, United Kingdom and New York, NY, USA.

McKay, NP & Kaufman, DS 2014, 'An extended Arctic proxy temperature database for the past 2,000 years', *Scientific Data*, vol. 1, p. 140026.

McIntyre, S & McKitrick, R 2003, 'Corrections to the Mann et. al. (1998) Proxy Data Base and Northern Hemispheric Average Temperature Series', *Energy & Environment*, vol. 14, no. 6, pp. 751–771.

McIntyre, S & McKitrick, R 2009, 'Proxy inconsistency and other problems in millennial paleoclimate reconstructions', *Proceedings of the National Academy of Sciences of the United States of America*, vol. 106, no. 6, E10, author reply E11.

Miller, DR, Habicht, MH, Keisling, BA, Castañeda, IS & Bradley, RS 2018, 'A 900-year New England temperature reconstruction from in situ seasonally produced branched glycerol dialkyl glycerol tetraethers (brGDGTs)', *Climate of the Past*, vol. 14, pp. 1653–1667.

Moberg, A, Sonechkin, DM, Holmgren, K, Datsenko, NM, Karlen, W & Lauritzen, SE 2005, 'Highly variable Northern Hemisphere temperatures reconstructed from low- and high-resolution proxy data', *Nature*, vol. 433, no. 7026, pp. 613–617.

Mokhova, II & Smirnov, DA 2018, 'Estimating the Contributions of the Atlantic Multidecadal Oscillation and Variations in the Atmospheric Concentration of Greenhouse Gases to Surface Air Temperature Trends from Observations', *Doklady Earth Sciences*, vol. 480, no. 1, pp. 602–606.

National Research Council 2006, 'Surface Temperature Reconstructions for the Last 2,000 Years', Washington, DC: The National Academies Press.

National Oceanic and Oceanic Administration, National Centres for Environmental Information. Paleoclimatology Data, viewed 26 February 2024, https://www.ncdc.noaa.gov/data-access/paleoclimatology-data

North, GR 2012, 'Apportioning natural and forced components in climate change', *Proceedings of the National Academy of Sciences of the United States of America*, vol. 109, no. 36, pp. 14285–14286.

Ortega, P, Lehner, F, Swingedouw, D, Masson-Delmotte, V, Raible, CC, Casado, M & Yiou, P 2015, 'A model-tested North Atlantic Oscillation reconstruction for the past millennium', *Nature*, vol. 523, pp. 71–74.

Parker, DE, Legg, TP & Folland, CK 1992, 'A new daily central England temperature series, 1772–1991', *International Journal of Climatology*, vol. 12, pp. 317–342.

Pei, Q, Zhang, DD, Li, J & Fei, J 2019, 'Proxy-based temperature reconstruction in China for the Holocene', *Quaternary International*, vol. 521, pp. 168–174.

Ramos-Román, MJ, Jiménez-Moreno, G, Camuera, J, García-Alix, A, Anderson, RS, Jiménez-Espejo, FJ, Sachse, D, Toney, J, Carrión, JS, Webster, CY & Yanes, C 2018, 'Millennial-scale cyclical environment and climate variability during the Holocene in the western Mediterranean region deduced from a new multiproxy analysis from the Padul record (Sierra Nevada, Spain)', *Global and Planetary Change*, vol. 168, pp. 35–53.

Raspopov, OM, Dergachev, VA, Esper, J, Kozyreva, OV, Frank, D, Ogurtsov, M, Kolström, T & Shao, X 2008, 'The influence of the de Vries (~200-year) solar cycle on climate variations: Results from the Central Asian Mountains and their global link', *Palaeogeography, Palaeoclimatology, Palaeoecology*, vol. 259, pp. 6–16.

Santer, BD, Painter, JF, Bonfils, C, Mears, CA, Solomon, S, Wigley, TML, Gleckler, PJ, et al. 2013, 'Human and natural influences on the changing thermal structure of the atmosphere', *Proceedings of the National Academy of Sciences of the United States of America*, vol. 110, no. 43, pp. 17235–17240.

Scafetta, N 2013, 'Discussion on climate oscillations: CMIP5 general circulation models versus a semi-empirical harmonic model based on astronomical cycles', *Earth-Science Reviews*, vol. 126, pp. 321–357.

Scafetta, N 2019,'On the reliability of computer-based climate models', *Italian Journal of Engineering Geology and Environment*, vol. 1, pp. 49–70.

Schneider, L, Smerdon, JE, Buntgen, U, Wilson, RJS, Myglan, VS, Kirdyanov, AV & Esper, J 2015, 'Revising mid-latitude summer temperatures back to A.D. 600 based on a wood density network', *Geophysical Research Letters*, vol. 42, pp. 4556–4562.

Schulz, M 2002, 'On the 1470-year pacing of Dansgaard–Oeschger warm events', *Paleoceanography*, vol. 17, no. 2, p. 1014.

Semenov, VA, Latif, M, Dommenget, D, Keenlyside, NS, Strehz, A, Martin, T & Park, W 2010, 'The impact of North Atlantic–Arctic multidecadal variability on Northern Hemisphere surface air temperature', *Journal of Climate*, vol. 23, pp. 5668–5677.

Soon, W & Baliunas, S, 2003. 'Proxy climatic and environmental changes of the past 1000 years', *Climate Research*, vol. 23, pp. 89–110.

Soon, W, Connolly, R & Connolly, M 2015, 'Re-evaluating the role of solar variability on Northern Hemisphere temperature trends since the 19th Century', *Earth Science Reviews*, vol. 150, pp. 409–452.

Suess, HE 1980, 'The radiocarbon record in tree rings of the last 8000 years', *Radiocarbon*, vol. 22, pp. 200–209, 280.

Swanson, KL, Sugihara, G, Tsonis, AA & May, R 2009, 'Long-Term Natural Variability and 20th Century Climate Change', *Proceedings of the National Academy of Sciences of the United States of America*, vol. 106, no. 38, pp. 16120–16123.

Tett, SFB, Betts, R, Crowley, TJ, Gregory, J, Johns, TC, Jones, A, Osborn, TJ, Ostrom, E, Roberts, DL & Woodage, MJ 2007, 'The impact of natural and anthropogenic forcings on climate and hydrology since 1550', *Climate Dynamics*, vol. 28, 3–34.

Trachsel, M, Grosjean, M, Larocque-Tobler, I, Schwikowski, M, Blass, A & Sturm, M 2010, 'Quantitative summer temperature reconstruction derived from a combined biogenic Si and chironomid record from varved sediments of Lake Silvaplana (south-eastern Swiss Alps) back to AD 1177', *Quaternary Science Reviews*, vol. 29, pp. 2719–2730.

Turney, CSM, Kershaw, AP, Clemens, SC, Branch, N, Moss, PT & Fifield, LK 2004, 'Millennial and orbital variations of El Niño/Southern Oscillation and high-latitude climate in the last glacial period', *Nature*, vol. 428, pp. 306–310.

Usoskin, IG & Kovaltsov, GA 2008, 'Cosmic rays and climate of the Earth: Possible connection', *Comptes Rendus Geoscience*, vol. 340, pp. 441–450.

Van der Bilt, WGM, D'Andrea, WJ, Werner, JP & Bakke, J 2019, 'Early Holocene temperature oscillations exceed amplitude of observed and projected warming in Svalbard lakes', *Geophysical Research Letters*, vol. 46, pp. 14,732–14,741.

Van Geel, B & Ziegler, PA 2013, 'IPCC Underestimates the Sun's role in Climate Change', *Energy & Environment*, vol. 24, no. 3–4, pp. 432–453.

Vitale, D & Bilancia, M 2013, 'Role of the natural and anthropogenic radiative forcings on global warming: evidence from cointegration – VECM analysis', *Environmental and Ecological Statistics*, vol. 20, pp. 413–444.

Von Storch, H, Zorita, E, Jones, JM, Dimitriev, Y, González-Rouco, F & Tett, SFB 2004, 'Reconstructing Past Climate from Noisy Data', *Science*, vol. 306, pp. 679–882.

Wahl, ER & Ammann, CM 2007, 'Robustness of the Mann, Bradley, Hughes reconstruction of Northern Hemisphere surface temperatures: Examination of criticisms based on the nature and processing of proxy climate evidence', *Climatic Change*, vol. 85, no. 1–2, pp. 33–69.

Wan, H, Zhang, X & Zwiers, F 2019, 'Human influence on Canadian temperatures', *Climate Dynamics*, vol. 52, pp. 479–494.

Wilson, R, Wiles, G, D'Arrigo, R et al. 2007, 'Cycles and shifts: 1,300 years of multi-decadal temperature variability in the Gulf of Alaska', *Climate Dynamics*, vol. 28, pp. 425–440.

Young, NE, Schweinsberg, AD, Briner, JP & Schaefer, JM 2015, 'Glacier maxima in Baffin Bay during the Medieval Warm Period coeval with Norse settlement', *Science Advances*, vol. 1, iss. 11.

Zhao, XH & Feng, XS 2015, 'Correlation between solar activity and the local temperature of Antarctica during the past 11,000 years', *Journal of Atmospheric and Solar-Terrestrial Physics*, vol. 122, pp. 26–33.

Zhao, X, Soon, W & Velasco Herrera, VM 2020, 'Evidence for Solar Modulation on the Millennial-Scale Climate Change of Earth', *Universe*, vol. 6, p. 153.

Zorita, E, González-Rouco, F & Legutke, S 2003, 'Testing the Mann et al. (1998) approach to paleoclimate reconstructions in the context of a 1000-year control simulation with the ECHO-G coupled climate model', *Journal of Climate*, vol. 16, pp. 1378–1390.

Zorita, E, Gonzalez-Rouco, JF, von Storch, H, Montavez, JP & Valero, F 2005, 'Natural and anthropogenic modes of surface temperature variations in the last thousand years', *Geophysical Research Letters*, vol. 32, L08707.

Zorita, E, González-Rouco, F & von Storch, H 2007, 'Comments on "Testing the fidelity of methods used in proxy-base reconstructions of past climate"', *Journal of Climate*, vol. 20, pp. 3693–3698.

2: Climate Cycles Determined from the Ice Edge in the Barents Sea

ArcticInfo (https://www.barentswatch.no/arcticinfo/) downloaded 240416

Astrup Jensen, A 2023, 'The time trend of the Arctic Sea ice extent Since 2007 no significant decline has been observed', *Science of Climate Change*, Vol 3.4, pp. 353–358, https://doi.org/10.53234/scc202310/23.

Charvátová, I & Hejda, P 2014, 'Responses of the basic cycles of 178.7 and 2402 yr in solar-terrestrial phenomena during the Holocene', *Pattern Recogn. Phys.*, vol. 2, pp. 21–26, http://doi.org/10.5194/prp-2-21-2014

Falk-Petersen, S, Hop, H, Budgell, WP et al. 2000, 'Physical and Ecological Processes in the Marginal Ice Zone of the Northern Barents Sea during the Summer Melt Periods', *Journal of Marine Systems*, vol. 27, pp. 131–159. https://doi.org/10.1016/S0924-7963(00)00064-6

Falk-Petersen, S, Pavlov, V, Cottier, F et al. 2015, 'At the Rainbow's End – Productivity Hotspots Due to Upwelling along Arctic Shelves', *Polar Biology*, vol. 38, pp. 5–11, https://doi.org/10.1007/s00300-014-1482-1

Jose, PD 1965, 'Sun's Motion and Sunspots', *The Astronomical Journal*, vol. 70, pp. 193–200.

Lamb, HH 1995, *Climate History and the Modern World*, 2d ed. Routledge, p. 433.

Løyning, TB, Dick, C, Goodwin, H et al. 2003, 'IACPO Informal Report', *World Climate Research Programme, Arctic Climate System Study, ACSYS Historical Ice Chart Archive (1553–2002)*, p. 8.

The Norwegian Polar Institute Library Yearbook 1999.

Mörner, NA, Solheim, J-E, Humlum, O & Falk-Petersen, S 2020, 'Changes in Barents Sea ice Edge Positions in the Last 440 years: A Review of Possible Driving Forces', *International Journal of Astronomy and Astrophysics*, vol. 10, pp. 97–164, https://doi.org/10.4236/ijaa.2020.102008

Solheim, J-E, Falk-Petersen, S, Humlum, O & Mörner, NA 2021, 'Changes in Barents Sea Ice Edge Positions in the Last 442 Years, Part 2: Sun, Moon and Planets', *International Journal of Astronomy and Astrophysics*, 2021, vol. 11, pp. 279–341, https://doi.org/10.4236/ijaa.2021.112015

Vinje, T 1999, 'Barents Sea ice edge variation over the past 400 years', extended abstract, Workshop on Sea-Ice Charts of the Arctic, Seattle, WA. *World Meteorological Organization, WMO/ TD* no. 949, pp. 4–6.

Vinje, T 2001, 'Anomalies and Trends of Sea-Ice Extent and Atmospheric Circulation in the Nordic Seas during the Period 1864–1998', *Journal of Climate*, vol. 14, pp. 255–267.

Yndestad, H & Solheim, J-E 2017, 'The influence of solar system oscillations on the variability of the total solar irradiance', *New Astronomy*, vol. 51, pp. 135–152, http://dx.doi.org/10.1016/j.newast.2016.08.020

3: Using Artificial Intelligence to Forecast Rainfall, Separation of Natural and Anthropogenic Components of Climate Change, and to Forecast Cyclone Trajectory and Intensity

Abbot, J 2021, 'Using Oscillatory Processes in Northern Hemisphere Proxy Temperature Records to Forecast Industrial-era Temperatures', *Earth Sciences*, vol. 10, no. 3, pp. 95–117.

Abbot, J & Marohasy, J 2012, 'Application of artificial neural networks to rainfall forecasting in Queensland, Australia', *Advances in Atmospheric Science*, vol. 29, no. 4, pp. 717–730.

Abbot, J & Marohasy, J 2013, 'The potential benefits of using artificial intelligence for monthly rainfall forecasting for the Bowen Basin, Queensland, Australia', *Water Resources Management, VII. WIT Transactions on Ecology and the Environment*, vol. 171, pp. 287–297.

Abbot, J & Marohasy, J 2014, 'Input selection and optimisation for monthly rainfall forecasting in Queensland, Australia using artificial neural networks', *Atmospheric Research*, vol. 138, pp. 166–178.

Abbot, J & Marohasy, J 2015a, 'Forecasting of Monthly Rainfall in the Murray Darling Basin, Australia: Miles as a Case Study', *River Basin Management*, vol. VIII, pp. 149–159.

Abbot, J & Marohasy, J 2015b, 'Improving monthly rainfall forecasts using artificial neural networks and single-month optimisation: a case study of the Brisbane Catchment, Queensland, Australia', *Water Resources Management*, vol. VIII, pp. 3–13.

Abbot J, Marohasy J 2015c, 'Using lagged and forecast climate indices with artificial intelligence to predict monthly rainfall in the Brisbane Catchment, Queensland, Australia', *International Journal of Sustainable Development and Planning*, vol. 10, no. 1, pp. 29–41.

Abbot, J & Marohasy, J 2015d, 'Using artificial intelligence to forecast monthly rainfall under present and future climates for the Bowen Basin, Queensland, Australia', *International Journal of Sustainable Development and Planning*, vol. 10, no. 1, pp. 66–75.

Abbot, J & Marohasy, J 2016a, 'Forecasting monthly rainfall in the Bowen Basin of Queensland, Australia, using neural networks with Niño Indices for El Niño-Southern Oscillation', *Lecture Notes in Artificial Intelligence*, vol. 9992, pp. 88–100.

Abbot, J & Marohasy, J 2016b, 'Forecasting monthly rainfall in the Western Australian wheat-belt up to 18 months in advance using artificial neural networks', *Lecture Notes in Artificial Intelligence*, vol. 9992, pp. 71–87.

Abbot, J & Marohasy, J 2017a, 'Forecasting extreme monthly rainfall events in regions of Queensland, Australia using artificial neural networks', *International Journal Sustainable Development and Planning*, vol. 12, no. 7, pp. 1117–1131.

Abbot, J & Marohasy, J 2017b, 'Skilful rainfall forecasts from artificial neural networks with long duration series and single-month optimization', *Atmospheric Research*, vol. 197, pp. 289–299.

Abbot, J & Marohasy, J 2017c, 'The application of machine learning for evaluating anthropogenic versus natural climate change', *GeoResJ*, vol. 14, pp. 36-46.

REFERENCES

Abbot, J & Marohasy, J 2018, 'Forecasting of Medium-term Rainfall Using Artificial Neural Networks: Case Studies from Eastern Australia', Chapter 3, *Engineering and Mathematical Topics in Rainfall*, InTech Publishing.

Andres, HJ & Peltier, WR 2015, 'Attributing observed Greenland responses to natural and anthropogenic climate forcings', *Climate Dynamics*, vol. 45, pp. 2919–2936.

Andrews, T, Gregory, JM, Webb, MJ & Taylor, KE 2012, 'Forcing, feedbacks and climate sensitivity in CMIP5 coupled atmosphere-ocean climate models', *Geophysical Research Letters*, vol. 39, p. L09712.

Arrhenius, S 1896, 'On the influence of carbonic acid in the air upon the temperature of the ground', *Philosopher's Magazine,* vol. 41, no. 5, pp. 237–276.

Chen, G, Chen, Z, Zhou, F, Zhang, H & Zhu, L 2019, 'A Semisupervised Deep Learning Framework for Tropical Cyclone Intensity Estimation', 10th International Workshop on the Analysis of Multitemporal Remote Sensing Images (MultiTemp), 8866970.

Chen, S, Wang, X, Chen, R, Lin, Q & Zhang, W 2022, 'TEMPO-RI: A Multi-Task Spatio-Temporal Model for Tropical Cyclone Rapid Intensification Forecasting', IEEE Smart World, Ubiquitous Intelligence and Computing, Autonomous and Trusted Vehicles, Scalable Computing and Communications, Digital Twin, Privacy Computing, Metaverse, pp. 864–871.

Chen, BF, Kuo, YT, Huang, TS, 2023, 'A deep learning ensemble approach for predicting tropical cyclone rapid intensification', *Atmospheric Science Letters*, vol. 24, no. 5, e1151.

Cheung, HM, Ho, C-H, Chang, M, Kim, J & Choi, W 2021, 'Development of a Track-Pattern-Based Medium-Range Tropical Cyclone Forecasting System for the Western North Pacific', *Weather and Forecasting*, vol. 36, no 4, pp. 1505–1518.

Chylek, P, Klett, JD, Dubey, MK & Hengartner, N 2016, 'The role of Atlantic Multi-decadal Oscillation in the global mean temperature variability', *Climate Dynamics,* vol. 47, pp. 3271–3279.

Esper, J & Cook, ER, Schweingruber, FH 2002, 'Low-Frequency Signals in Long Tree-Ring Chronologies for Reconstructing Past Temperature Variability', *Science*, vol. 295, p. 5563.

Flato, G, Marotzke, J, Abiodun, B, Braconnot, P, Chou, SC, Collins, W, Cox, P, Driouech, F, Emori, S, Eyring, V, Forest, C, Gleckler, P, Guilyardi, E, Jakob, C, Kattsov, V, Reason, C & Rummukainen, M 2013, *'Evaluation of climate models', Climate change 2013: the physical science basis. Contribution of working group I to the Fifth Assessment Report of the Intergovernmental Panel on Climate Change*, Cambridge, United Kingdom and New York, NY, USA. Cambridge University Press.

Geng, X, Liu, Z & Shi, Z 2023, 'Spatio-Temporal Alignment and Track-To-Velocity Module for Tropical Cyclone Forecast', *Remote Sensing*, vol. 15, no. 20, p. 4938.

Gervais, F 2016, 'Anthropogenic CO_2 warming challenged by 60-year cycle', *Earth-Science Reviews*, vol. 155, pp. 129–135.

Guangliang, H, Chongyi, E, Xiangjun, L & Fangming, Z 2013, 'Reconstruction of integrated temperature series of the past 2000 years on the Tibetan plateau with 10-year intervals', *Theoretical and Applied Climatology*, vol. 113, pp. 259–269.

Hunt, BG 2006, 'The medieval warm period, the little ice age and simulated climatic variability', *Climate Dynamics*, vol. 2 no. 7, pp. 677–694.

IPCC 1990, Houghton, JT et al. (eds.) 1990, *Assessment Report prepared for the Intergovernmental Panel on Climate Change by Working Group 1*, Cambridge, England, Cambridge University Press.

Ko, MC, Chen, X, Kubat, M & Gopalakrishnan, S 2023, 'The Development of a Consensus Machine Learning Model for Hurricane Rapid Intensification Forecasts with Hurricane Weather Research and Forecasting (HWRF) Data', *Weather and Forecasting*, vol. 38, no. 8, pp. 1253–1270.

Kolukula, SS & Murty, PLN 2022, 'Improving cyclone wind fields using deep convolutional neural networks and their application in extreme events', *Progress in Oceanography*, vol. 202, 102763.

Kovordányi, R & Roy, C 2009, 'Cyclone track forecasting based on satellite images using artificial neural networks', *ISPRS Journal of Photogrammetry and Remote Sensing*, vol. 64, no. 6, pp. 513–521.

Lam, R, Sanchez-Gonzalez, A, Willson, M, Wirnsberger, P, Fortunato, M, Alet, F, Ravuri, S, Ewalds, T, Eaton-Rosen, Z, Hu, W, Merose, A, Hoyer, S, Holland, G, Vinyals, O, Stott, J, Pritzel, A, Mohamed, S & Battaglia, P 2023, 'Learning skillful medium-range global weather forecasting', *Science*, https://doi.org/10.1126/science.adi2336

Lamb, HH 1965, 'The earlier medieval warm epoch and its sequel', *Palaeogeography, Palaeoclimatology, Palaeoecology*, vol. 1, pp. 13–37.

Lamb, HH 1982, *Climate, history and the modern world*, New York, Routledge.

Lin, Q, Jin, Y, Lin, Y & Li, X 2023, 'Three-dimensional temporal-spatial attention for tropical cyclone forecast', *Proceedings of SPIE – The International Society for Optical Engineering*, 12918, 1291820.

Ljungqvist, FC, Krusic, PJ, Brattstrom, G, Sundqvist, HS 2012, 'Northern Hemisphere temperature patterns in the last 12 centuries', *Climate of the Past*, vol. 8, pp. 227–49.

Lüdecke, HJ, Hempelmann, A & Weiss, CO 2013, 'Multi-proxy climate dynamics: spectral analysis of long-term instrumental and proxy temperature records', *Climate of the Past*, vol. 9, pp. 447–552.

Lüdecke, HJ & A, Weiss, CO 2017, 'Harmonic Analysis of Worldwide Temperature Proxies for 2000 Years', *The Open Atmospheric Science Journal*, vol. 17, pp. 44–53.

Mann, ME, Zhang, Z, Hughes, MK, Bradley, RS, Miller, SK, Rutherford, S & Ni, FB 2008, 'Proxy-based reconstructions of hemispheric and global surface temperature variations over the past two millennia', *Proceedings of the National Academy of Sciences*, vol. 105, no. 36, pp. 13252–13257.

McShane, BB & Wyner, AJ 2011, 'A statistical analysis of multiple temperature proxies: are reconstructions of surface temperatures over the last 1000 years reliable?', *Annals of Applied Statistics*, vol. 5, no.1, pp. 5–44.

Moberg, A 2013, 'Comparisons of simulated and observed Northern Hemisphere temperature variations during the past millennium – selected lessons learned and problems encountered', *Tellus* B, vol. 65, p. 19921.

Myhre, G, Shindell, D, Bréon, FM, Collins, W, Fuglestvedt, J, Huang, J et al. 2013, *Anthropogenic and natural radiative forcing Climate change 2013: The physical science basis*, Contribution of working group I to the Fifth Assessment Report of the Intergovernmental Panel On Climate Change, Cambridge, United Kingdom and New York, NY, USA. Cambridge University Press.

Na, Y, Na, B & Son, S 2022, 'Near real-time predictions of tropical cyclone trajectory and intensity in the northwestern Pacific Ocean using echo state network', *Climate Dynamics*, vol. 58, no. 3–4, pp. 651–667.

Neukom, R, Luterbacher, J, Villalba, R, Kuttel, M, Frank, D, Jones, PD, Grosjean, M, Wanner, H, Aravena, JC, Black, DE, Christie, DA, D'Arrigo, R, Lara, A, Morales, M, Soliz-Gamboa, C, Srur, A, Urrutia, R & von Gunten, L 2011, 'Multiproxy summer and winter surface air temperature field reconstructions for southern South America covering the past centuries', *Climate Dynamics*, vol. 37, no. 1-2, pp. 35–51.

Priem, HNA 1997, 'CO$_2$ and climate: A geologist's view', *Space Science Reviews*, vol. 81, pp. 173–198.

Roy, C & Kovordányi, R 2012, 'Tropical cyclone track forecasting techniques – A review', *Atmospheric Research*, vol. 104, pp. 40–69.

Sahoo, B & Bhaskaran, PK 2019, 'Prediction of storm surge and inundation using climatological datasets for the Indian coast using soft computing techniques', *Soft Computing*, vol. 23, pp. 12363–12383.

Sarkar, I, Chaudhuri, S & Pal, J 2022, 'Forecasting pressure drop and maximum sustained wind speed associated with cyclonic systems over Bay of Bengal with neuro-computing', *Theoretical and Applied Climatology*, vol. 149, pp. 1255–1276.

Scafetta, N 2013, 'Discussion on climate oscillations: CMIP5 general circulation models versus a semi-empirical harmonic model based on astronomical cycles', *Earth-Science Reviews*, vol. 126, pp. 321–357.

Snaiki, R & Wu, T 2022, 'Knowledge-Enhanced Deep Learning for Simulation of Extratropical Cyclone Wind Risk', *Atmosphere*, vol. 13, no. 5, p. 757.

Tan, J, Chen, S & Wang, J 2021, 'Western North Pacific tropical cyclone track forecasts by a machine learning model', *Stochastic Environmental Research and Risk Assessment*, vol. 35, no. 6, pp. 1113–1126.

Taricco, C, Mancuso, S, Ljungqvist, FC, Alessio, S & Ghil, M 2015, 'Multispectral analysis of Northern Hemisphere temperature records over the last five millennia', *Climate Dynamics*, vol. 45, pp. 83–104.

van Geel, B & Ziegler, PA 2013, 'IPCC Underestimates the Sun's role in climate change', *Energy and the Environment*, vol. 24, no. 3-4, pp. 431–453.

Wan, H, Zhang, X & Zwiers, F 2019, 'Human influence on Canadian temperatures', *Climate Dynamics*, vol. 52, pp. 479–494.

Wang, L, Wan, B, Zhou, S, Sun, H & Gao, Z 2023, 'Forecasting tropical cyclone tracks in the northwestern Pacific based on a deep-learning model', *Geoscientific Model Development*, vol. 16, no. 8, pp. 2167–2179.

Wu, Y, Geng, X, Liu, Z & Shi, Z 2022, 'Tropical Cyclone Forecast Using Multitask Deep Learning Framework', *IEEE Geoscience and Remote Sensing Letters*, 19.

4: The Geology of Climate Change

Alley, RB 2000, 'The Younger Dryas cold interval as viewed from central Greenland', *Quaternary Science Reviews,* vol. 19, pp. 213–226.

Allison, CM, Roggensack, K & Clarke, AV 2019, 'H_2O-CO_2 solubility in alkali-rich mafic magmas: new experiments at mid-crustal pressures', *Contributions to Mineralogy and Petrology,* vol. 174, no. 58.

Bernier, RA & Kothavala, Z 2001, 'GEOCARBIII: A revised model of atmospheric CO_2 over Phanerozoic time', *American Journal of Science,* vol. 301, pp. 182–204.

Braun, H, Christi, M, Rahmstorf, S, Ganopolski, A, Mangini, A, Kubatzki, C, Roth, K & Kromer, B 2005, 'Possible solar origin of the 1,470-year glacial cycle demonstrated in a coupled model', *Nature,* vol. 438, no. 7065.

Bufe, A, Hovius, N, Emberson, R, Rugenstein, JKC, Galy, A, Hassenruck-Gudipati, HJ & Chang, J-M 2021, 'Co-variation of silicate, carbonate and sulfide weathering drives CO_2 release with erosion', *Nature Geoscience,* vol. 14, pp. 211–216.

Burton-Johnson, A, Dziadek, R & Martin, C 2020, 'Review article: Geothermal heat flow in Antarctica: current and future directions', *The Cryosphere,* vol. 14, no. 11, pp. 3843–3873.

Catling, DC & Zahnle, KJ 2020, 'The Archean atmosphere', *Science Advances,* vol. 6, no. 9.

Corrocedo, JC, Troll, VR, Zoczek, K, Rodríguez-González, A, Soler, V & Deegon, FM 2015, 'The 2011-2012 submarine eruption off El Hierro, Canary Islands: New lessons in oceanic island growth and volcanic crisis management'. *Earth-Science Reviews,* vol. 150, pp. 168–200.

Day, A 2020, 'Fire and ice, volcanoes at Antarctica'. In J. Marohasy (ed.) *Climate Change: The facts 2020*, ch. 4, Institute of Public Affairs, pp, 61–80.

Deegan, FM, Cellegaro, S, Davies, JHFL & Svensen, HH 2023, 'Driving global change one LIP at a time', *Elements,* vol. 19, pp. 269–275.

Dixon, JE, Stolper, EM & Holloway, JR 1995, 'An experimental study of water and carbon dioxide solubilities in mid-ocean ridge basaltic liquids. Part 1: Calibration and solubility models', *Journal of Petrology,* vol. 36, iss. 6, pp. 1607–1631.

Dumont, S, Custódio, S, Petrosine, S, Thomas, AM & Sottili, G 2023, 'Tides, earthquakes and volcanic eruptions'. In M.Green & JC Duarte (eds) *A journey through tides*, ch. 14, Elsevier, pp. 333–364.

Eguchi, J, Diamond, CW & Lyons, TW 2022: 'Proterozoic supercontinent-break-up as a driver for oxygenation and subsequent carbon isotope excursions', *Proceedings of the National Academy of Sciences Nexus*, vol. 1, no. 2.

Evangelinos, D, Etourneau, J, van de Flierdt, T, Crosta, X, Jeandel, C, Flores, J-A, Harwood, DM, Valero, L, Ducassou, E, Sauermilch, I, Klocker, A, Cacho, I, Pena, LD, Kreissig, K, Benoit, M, Belhadi, M, Paredes, E, Garcia-Solsona, E, López-Quirós, A, Salabarnada, A & Escutia, C 2024, 'Late Miocene onset of the modern Antarctic Circumpolar Current', *Nature Geoscience,* vol. 17, pp. 165–170.

Feynman, J 1982, 'Geomagnetic and solar wind cycle', 1900–1975, *Journal of Geophysical Research,* vol. 87, no. A8, pp. 6153–6162.

Fuchs, S, Norden, B & International Heat Flow Commission 2021, 'The global heat flow database. Release 2021', *GFZ Data Services.*

Gerhart, LM & Ward, JK 2010, 'Plant responses to low [CO_2] of the past', *New Phytologist*, vol. 186. no. 3.

German, CR & von Damm, KL 2003, 'Hydrothermal processes', in H. Elderfield (ed.) *Treatise on Geochemistry*, vol 6, Elsevier, p. 181–222.

Gies, DR & Helsel, JW 2005, 'Ice age epochs and the Sun's path through the galaxy', *Astronomical Journal,* vol. 626, pp. 844–848.

Gillman, M & Erenler, H 2019, 'Reconciling the Earth's stratigraphic record with the structure of our galaxy', *Geoscience Frontiers,* vol. 10, no. 6, pp. 2147–2151.

Grasby, SE & Bond, DPG 2023, 'How large igneous provinces killed most life on Earth-numerous times', *Elements*, vol. 19, pp. 276-281.

Hamza, VM & Viera, FP 2012, 'Global distribution of the lithosphere-asthenosphere boundary: A new look', *Solid Earth Discussions*, vol. 4, pp. 279–313.

Hoffman, PF, Abbot, DS, Ashkenazy, Y, Benn, DI, Brooks, JJ, Cohen, PA, Cox, GM, Creveling, JR, Donnadieu, Y, Erwin, DH, Fairchild, IJ, Ferreira, D, Goodman, JC, Halverson, GP, Jansen, MF, Le Hir, G, Love, GD, Macdonald, FA, Maloof, AC, Partin, CA, Ramstein, G, Rose, BEJ, Rose, CV, Sadier, PM, Tziperman, E, Voigt, A & Warren, SG 2017: 'Snowball Earth climate dynamics and Cryogenian geology-geobiology', *Science Advances*, vol. 3, no. 11.

joidesresolution.org, Seawater circulation and ocean composition, 14 October 2011, viewed 21 February 2024.

Kaufman, AJ & Xiao, S 2003, 'High CO_2 levels in the Proterozoic atmosphere estimated from analyses of individual microfossils', *Nature*, vol. 425, no. 6955, pp. 279–283.

Large, RR, Mukherjee, I, Gregory, D, Steadman, J, Corkrey, R & Danyushevsky, LV 2019, 'Atmosphere oxygen cycling through the Proterozoic and Phanerozoic', *Mineralium Deposita*, vol. 54, pp. 485–506.

Lee, D & Veizer, J 2003, 'Water and carbon cycles in the Mississippi river basin: potential implications for the northern hemisphere "residual terrestrial sink"', *Global Biogeochemical Cycles*, vol. 17, no. 2.

Li, Y, Peng, S, Liu, H, Zhong, X, Ye, W, Han, Q, Zhong, Y, Xu, L & Li, Y 2020, 'Westerly jet stream controlled climate change mode since the Last Glacial Maximum in the northern Qinghai-Tibet Plateau', *Earth and Planetary Science Letters*, vol. 549.

Lindzen, RS 1997, 'Can increasing carbon dioxide cause climate change'? *Proceeding of the National Academy of Sciences*, vol. 94, pp. 8335–8342.

Lockwood, M 2007, 'What do cosmogenic isotopes tell us about past solar forcing of climate?', *Space Science Reviews*, vol. 125, pp. 95–109.

Loehle, C & Scafetta, N 2012, 'Climate change attribution using empirical decomposition of climatic data', *The Open Atmospheric Science Journal*, vol. 5, pp. 74–86.

Lüdecke, H-J, Weiss, CO & Hempelmann, A 2015, 'Paleoclimate forcing by the solar De Vries/Suess cycle', *Climate of the Past*, vol. 11, pp. 279–305.

Macdonald, FA & Swanson-Hysell, NL, 2023, 'The Franklin Large Igneous Province and Snowball Earth initiation', *Elements*, vol. 19, pp. 296–301.

McConnell, JR, Burke, A, Dunbar, NW & Winkler, G 2017, 'Synchronous volcanic eruptions and abrupt climate change ~17.7ka plausibly linked by stratospheric ozone depletion', *Proceedings of the National Academy of Sciences*, vol. 114, no. 38, pp. 10035–10040.

Millán, L, Santee, ML, Lambert, A, Liversey, NJ, Werner, F, Schwartz, MJ, Pumphrey, HC, Manney, GL, Wang, Y, Su, H, Wu, L, Read, WG & Froidevaux, L 2023, 'The Hunga Tonga-Hunga Ha'apai hydration of the stratosphere,' *Geophysical Research Letters*, vol. 49, iss. 13.

Miyahara, H, Tokanai, F, Moriya, T, Takeyama, M, Sakurai, Horiuchi, K & Hotta, 2021, 'Gradual onset if the Maunder Minimum revealed by high-precision carbon-14 analyses', *Scientific Reports*, vol. 11, no. 1, p. 5482.

Muller, RA & MacDonald, GJ 1997, 'Spectrum of 100-kyr glacial cycle: orbital inclination not eccentricity', *Proceedings of the National Academy of Sciences*, vol. 94, no. 16, pp. 8329–8334.

Niu, Y 2020, 'On the cause of continental breakup: A simple analysis in terms of driving mechanisms of plate tectonics and mantle plumes', *Journal of Asian Earth Sciences*, vol. 194, p. 104367.

Pavlov, AA, Hurtgen, M, Kasling, JF & Arthur, MA 2003, 'Methane-rich Proterozoic atmosphere', *Geology*, vol. 31, no. 1.

Petit, JR, Jouzel, J, Raynaud, D, Barkov, NI, Barnola, J-M, Basile, I, Bender, M, Chappellaz, J, Davis, M, Delaygue, G, Delmotte, M, Kotlyakov, VM, Legrand, M, Lipenkov, VY, Lorius, C, Pépin, L, Ritz, Saltzman, E & Stievenard, M 1999, 'Climate and atmospheric history of the past 420,000 years from the Vostok ice core, Antarctica', *Nature*, vol. 399, no. 6735, pp. 429–436.

Rampino, MR & Self, S 1984, 'Sulphur-rich volcanic eruptions and stratospheric aerosols', *Nature*, vol. 310, pp. 677–679.

Rampino, MR, Strothers, RB & Self, S 1985, 'Climatic effects of volcanic eruptions', *Nature*, vol. 313, no. 272, pp. 1283–1300.

Sage, RF 2003, 'The evolution of C_4 photosynthesis', *New Phytologist*, vol. 161, no. 2.

Shaviv, NJ & Veizer, J 2003, 'Celestial driver of Phanerozoic climate?', *GSA Today*, vol. 13, no. 7, pp. 4–19.

Somoza, K, González, J, Barler, SJ, Mandureira, P, Medialdea, T, Ignacio, C de, Lourenço, N, Vázquez, JT & Palomino, D 2017, 'Evolution of submarine eruptive activity during the 2011–2012 El Hierro event as documented by hydroacoustic images and remotely operated vehicle observations', *Geochemistry, Geophysics, Geosystems*, vol. 18, no. 8, pp. 3109–3137.

Stål, T, Reading, AM, Fuchs, S, Halpin, JA & Turner, RJ 2022, 'Properties and biases of the global heat flow compilation', *Frontiers in Earth Science*, vol. 10.

Svensmark, H & Friis-Christiansen, E 1997, 'Variation of cosmic ray flux and global cloud coverage: A missing link in solar-climate relationships', *Journal of Atmospheric and Solar Terrestrial Physics*, vol. 59, pp. 1225–1232.

Svensmark, H 2000, 'Cosmic rays and Earth's climate', *Space Science Reviews*, vol. 93, pp.175–185.

Svensmark, H, Enghoff, MB, Shaviv, NJ & Svensmark, J 2017, 'Increased ionization supports growth of aerosols into cloud condensation nuclei', *Nature Communications*, vol. 8, no. 2199.

Thordarson, T & Self, S 2003, 'Atmospheric and environmental effects of the 1783–1784 Laki eruption: A review and reassessment', *Journal of Geophysical Research*, vol. 108, iss. D-1.

Tuttle, GF & Bowen, NL 1958, 'Origin of granite in the lights of experimental studies in the system $NaAlSi_3O_8$-$KAlSi_3O_8$-SiO_2-H_2O', Memoir 74, *Geological Society of America*.

Usoskin, IG, Gallet, Y, Lopes, F, Kovaltsov, GA and Hulot, G 2016, 'Solar activity during the Holocene: the Hallstatt cycle and its consequence for grand minima and maxima', *Astronomy and Astrophysics*, vol. 587, A150.

Usoskin, IG, Solanki, SK, Krivova, N, Hofer, B, Kovaltsov, GA, Wacker, L, Brehm, N & Kromer, B 2021, 'Solar cycle activity over the last millennium reconstructed from annual ^{14}C data', *Astronomy and Astrophysics*, vol. 649, A141.

Van den Ende, C, White, LT & van Welzen, PC 2017, 'The existence and break-up of the Antarctic land bridge as indicated by both amphi-Pacific distributions and tectonics', *Gondwana Research*, vol. 33, pp. 219–237.

Veizer, J, Godderis, Y & François, LM 2000, 'Evidence for decoupling of atmospheric CO_2 and global climate during the Phanerozoic era', *Nature*, vol. 408, pp. 698–701.

Wei-Ping, J, E, D-C, Zhan, B-W & Liu, Y-W 2009, 'New model of Antarctic plate motion and its analysis', *Chinese Journal of Geophysics*, vol. 52, no. 1, pp. 23–32.

Wilcock, WSD, Dziak, RP, Tolstoy, M, Chadwick, WW, Nooner, SL, Bohnenstiehl, DR, Caplan-Auerbach, J, Waldhauser, F, Arnulf, AF, Baillard, C, Lau, T-K, Haxel, JH, Tan, YJ, Garcia, C, Levy, S & Mann, ME 2018, 'The recent volcanic history of Axial Seamount: Geophysical insights into past eruption dynamics with an eye toward enhanced observations of future eruption', *Oceanography*, vol. 31, no. 1. pp. 114–123.

Yim, W 2022 'Volcanic eruptions, a driver of natural climate variability – ignored by IPCC', *Hong Kong Institution of Engineers*.

Zhang, T, Li, M, Chen, X, Wang, T & Shen, Y 2022, 'High atmospheric CO_2 levels in the early Mesoproterozoic estimated from paired carbon isotopic records from carbonates from North China', *Precambrian Research*, vol. 380, no. 106812.

Zharkova, VV, Vasilieva, I, Shepherd, S & Popova, E 2023, 'Periodicities of solar activity and solar radiation derived from observations and their links with the terrestrial environment', *Natural Science,* vol. 15, no. 3.

5: Why Climate Models Fail

Bony, S, Colman, R, Kattsov, VM, Allan, RP, Bretherton, CS, Dufresne, J-L et al. 2006, 'How well do we understand and evaluate climate change feedback processes?' *Journal of Climate*, vol. 19, pp. 3445–3482.

British Petroleum (BP) 2019, Statistical Review of World Energy, https://www.bp.com/en/global/corporate/news-and-insights/press-releases/bp-statistical-review-of-world-energy-2019.html

Butler, JH & Montzka, SA 2021, The NOAA annual greenhouse-gas index (AGGI), Global Monitoring Laboratory, National Oceanographic and Atmospheric Administration, Boulder, Colorado, 2021, viewed June 2021.

Christy, JR, Spencer, RW, Braswell, WD & Junod, R 2018, 'Examination of space-based bulk atmospheric temperatures used in climate research', *International Journal of Remote Sensing*, vol. 39, iss. 11, pp. 3580–3607.

Environmental Protection Agency (EPA) 2023, Global Greenhouse Gas Emissions, https://www.epa.gov/ghgemissions/global-greenhouse-gas-emissions-data

ESRL 2023, 'Trend in specific humidity at 300 mb', Earth System Research Laboratory, National Oceanographic and Atmospheric Administration, reported in Humlum, O 2023, Monthly climate statistics, http://climate4you.com

France24.com 2022, China doubles down on coal as energy crunch bites, https://www.france24.com/en/live-news/20220918-china-doubles-down-on-coal-as-energy-crunch-bites

Frank, P 2019, 'Propagation of error and the reliability of global air temperature projections', *Frontiers in Earth Science*, vol. 7, p. 223.

Friedlingstein, P, O'Sullivan, M, Jones, MW & Andrew, RM 2022, 'Global carbon budget 2022', *Earth System Science Data*, vol. 14, iss. 11, pp. 4811–4900, https://essd.copernicus.org/articles/14/4811/2022

Globalpetrolprices.com 2023, Electricity prices by country, https://www.globalpetrolprices.com/electricity_prices

Grinsted, A, Moore, JC & Jevrejeva, S 2009, 'Reconstructing sea level from paleo and projected temperatures 200 to 2100 AD', *Climate Dynamics*, vol 34, pp. 461–472.

HadCRUT5: Morice, CP, Kennedy, JJ, Rayner, NA, Winn, JP, Hogan, E, Killick, RE et al. 2021, 'An updated assessment of near-surface temperature change from 1850: the HadCRUT5 dataset', *JGR Atmospheres*, vol. 126, iss. 3, https://doi.org/10.1029/2019JD032361, viewed May 2023 https://crudata.uea.ac.uk/cru/data/temperature/HadCRUT5.0Analysis_gl.txt

Hansen, J, Lacis, A, Rind, D & Russell, G 1984, 'Climate sensitivity: Analysis of Feedback Mechanisms', *Climate Processes and Climate Sensitivity* (AGU Geophysical Monograph 29), Hansen J & Takahashi T (eds.), *American Geophysical Union,* pp. 130–163.

IPCC 1990, Houghton, JT et al. (eds.) 1990, *Assessment Report prepared for the Intergovernmental Panel on Climate Change by Working Group 1*, Cambridge, England, Cambridge University Press.

IPCC, 2001, Houghton, JT et al. (eds.), *Climate Change 2001: The Scientific Basis. Contribution of Working Group I to the Third Assessment Report of the Intergovernmental Panel on Climate Change*, Cambridge, England, Cambridge University Press.

IPCC 2007, Solomon, S, et al. (eds.), *Climate Change 2007: The physical science basis. Contribution of Working Group I to the Fourth Assessment Report of the Intergovernmental Panel on Climate Change*, Cambridge, England, Cambridge University Press.

IPCC 2013, Stocker, TF et al. (eds.), *Climate change 2013: The physical science basis. Contribution of Working Group I to the Fifth Assessment Report of the Intergovernmental Panel on Climate Change*, Cambridge, England, Cambridge University Press.

IPCC 2021, Masson-Delmotte, V, et al. (eds), *Climate Change 2021: The physical science basis. Contribution of Working Group I to the Sixth Assessment Report of the Intergovernmental Panel on Climate Change*, Cambridge, England, Cambridge University Press.

Irving, D, Hobbs, W, Church, J & Zika, J 2021, 'A Mass and Energy Conservation Analysis of Drift in the CMIP6 Ensemble', *Journal of Climate*, vol. 34, iss. 8, pp.3157–3170.

Karl, TR, Hassol, SJ, Miller, CD & Murray, WL 2006, Temperature Trends in the Lower Atmosphere: Steps for Understanding and Reconciling Differences, U.S. Climate Change Science Program.

Kumra, G & Woetzel, J 2022, Cost to get to net-zero. McKinsey Global Institute, https://www.mckinsey.com/mgi/overview/in-the-news/what-it-will-cost-to-get-to-net-zero

Lorenz, EN 1963, 'Deterministic Nonperiodic Flow', *Journal of Atmospheric Sciences*, vol. 20, iss. 2, pp. 130–141.

McKitrick, R & Christy, J 2020, 'Pervasive Warming Bias in CMIP6 Tropospheric Layers', *Earth & Space Science*, vol. 7, iss. 9.

Meinshausen, M, Vogel, E, Nauels, A, Lorbacher, K, Meinshausen, N, Etheridge, DM, Fraser, PJ et al. 2017, 'Historical greenhouse gas concentrations for climate modelling (CMIP6)', *Geosciences Model Development*, vol. 10, pp. 2057–2116.

National Grid ESO 2020, Analysing the costs of our Future Energy Scenarios, London, https://www.nationalgrideso.com/news/analysing-costs-our-future-energy-scenarios

Global Monitoring Laboratory 2023, The NOAA Annual Greenhouse Gas Index (AGGI), Fig. 3, viewed 23 October 2023, https://gml.noaa.gov/aggi/aggi.html

Office for National Statistics (ONS) 2021, UK greenhouse gas emissions, provisional figures, Department for Business, Energy & Industrial Strategy, London.

Paltridge, G, Arking, A & Pook, M 2009, 'Trends in middle- and upper-level tropospheric humidity from NCEP reanalysis data', *Theoretical and Applied Climatology*, vol. 98, pp. 351–359.

Pinker, RT, Zhan, B & Dutton, EG 2005, 'Can satellites observe trends in surface solar radiation?' *Science*, vol. 308, pp. 850–854.

Reuters.com 2022, Analysis: India power binges on coal, outpaces Asia, https://www.reuters.com/business/cop/india-power-binges-coal-outpaces-asia-2022-11-18

Reuters.com 2023, Pakistan plans to quadruple domestic coal-fired power, https://www.reuters.com/business/energy/pakistan-plans-quadruple-domestic-coal-fired-power-move-away-gas-2023-02-13

University of Alabama in Huntsville (UAH) 2023, Monthly global mean lower-troposphere temperature anomalies, viewed May 2023, http://www.nsstc.uah.edu/data/msu/v6.0/tlt/uahncdc_lt_6.0.txt

Vial, J, Dufresne, J-L & Bony, S 2013, 'Interpretation of inter-model spread in CMIP5 climate sensitivity estimates', *Climate Dynamics*, vol. 41, pp. 3339–3362.

6: Global Climate Lysenkoism: The Politics of Preferring Models Over Observations

Alimonti, G, Mariani, L, Prodi, F & Ricci, RA 2022, 'A critical assessment of extreme events trends in times of global warming', *The European Physical Journal Plus*, vol. 137, no. 1, pp. 1–20.

Arrhenius, S 1896, 'On the Influence of Carbonic Acid in the Air upon the Temperature of the Ground', *Philosophical Magazine and Journal of Science*, series 5, vol. 41, pp. 237–276.

Babyak, MA 2004, 'What you see may not be what you get: A brief, nontechnical introduction to overfitting in regression-type models', *Psychosomatic Medicine*, vol. 66, pp. 411–421.

Barrett, J 2005, 'Greenhouse molecules, their spectra and function in the atmosphere', *Energy & Environment*, vol. 16, no. 6, pp. 1037–1045.

Box, GE 1976, 'Science and statistics', *Journal of the American Statistical Association*, vol. 71, no. 356, pp. 791–799.

Bryce, R, 2023, 'The Anti-Industry Industry', visited 6 January 2024, https://robert bryce.substack.com/p/the-anti-industry-industry

Christy, JR 2016, U.S. House Committee on Science, Space & Technology, 2 February 2016, Testimony of John R. Christy, University of Alabama in Huntsville.

Coe, D, Fabinski, W & Wiegleb G 2021, 'The impact of CO_2 and other "Greenhouse Gases" on equilibrium Earth temperatures', *International Journal of Atmospheric and Oceanic Sciences*, vol. 15, no. 2.

Ellwanger, A 2023, 'Settled Science and the Politics of Knowledge', *The American Mind*, visited 6 January 2024, https://americanmind.org/salvo/settled-science-and-the-politics-of-knowledge/

Fall, S, Watts, A, Nielsen-Gammon, J, Jones, E, Niyogi, D, Christy, JR & Pielke Sr, RA 2011, 'Analysis of the impacts of station exposure on the US Historical Climatology Network temperatures and temperature trends', *Journal of Geophysical Research: Atmospheres*, vol. 116, no. D14.

Festinger, L 1957, 'A theory of cognitive dissonance', *Palo Alto*, Stanford Univ. Press, California.

Festinger, L, Riecken, HW & Schachter, S 1956, *When Prophecy Fails: a Social and Psychological Study of a Modern Group that Predicted the Destruction of the World*, Harper & Row, New York.

Freedman, DA 1991, 'Statistical models and shoe leather (with discussion)', *Sociological Methodology*, vol. 21, pp. 291–313.

Frické, M 2015, 'Big data and its epistemology', *Journal of the association for information science and technology*, vol. 66, no. 4, pp. 651–661.

Head, BW 2013, 'Evidence-based policymaking–speaking truth to power?' *Australian Journal of Public Administration*, vol. 72, no. 4, pp. 397–403.

Hull, DL 2019, *Science as a process: an evolutionary account of the social and conceptual development of science*, University of Chicago Press, Chicago.

Ioannidis, JP 2005, 'Why most published research findings are false', *PLoS medicine*, vol. 2, no. 8, e124.

Janis, IL 1972, *Victims of Groupthink*, Houghton Mifflin, Boston.

Jucker, M, Lucas, C & Dutta, D 2023, 'Long-term surface impact of Hunga Tonga-Hunga Ha'apai-like stratospheric water vapor injection,' *Authorea Preprints*, DOI: 10.22541/essoar.169111653.36341315/v1

Kellow, AJ 2007, *Science and public policy: the virtuous corruption of virtual environmental science*, Edward Elgar Publishing.

Kellow, A 2018, 'From Policy Typologies to Policy Feedback', in HK Colebatch, & R Hoppe, eds, *Handbook on Policy, Process and Governing*, Edward Elgar, Cheltenham, pp. 457–472.

Kellow, A 2020, 'The Lure of the Apocalypse', *Quadrant*, viewed 6 January 2024, https://quadrant.org.au/magazine/2020/01/the-lure-of-the-apocalypse/

Kerr, DH 1976, 'The logic of "policy" and successful policies', *Policy Sciences*, vol. 7, no. 3, pp. 351–363.

Khaykin, S, Podglajen, A, Ploeger, F, Grooß, JU, Tencé, F, Bekki, S, Khlopenkov, K, Bedka, K, Rieger, L, Baron, A & Godin-Beekmann, S 2022, 'Global perturbation of stratospheric water and aerosol burden by Hunga eruption,' *Communications Earth & Environment*, vol. 3, no. 1, p. 316.

Lewis, N 2023, 'Objectively combining climate sensitivity evidence,' *Climate Dynamics*, vol. 60, no. 9, pp. 3139–3165.

Lindblom, CE 1977, *Politics and Markets*, Basic Books, New York.

Lomborg, B 2020, 'Welfare in the 21st century: Increasing development, reducing inequality, the impact of climate change, and the cost of climate policies,' *Technological Forecasting and Social Change*, vol. 156, p. 119981.

Lowi, TJ 1964, 'American Business, Public Policy, Case Studies, and Political Theory,' *World Politics*, vol. 16, pp. 677–715.

McKibben, W 2022, 'From Climate Exhortation to Climate Execution', *The New Yorker*, 27 December, https://www.newyorker.com/news/daily-comment/from-climate-exhortation-to-climate-execution

Marohasy, J 2020, 'Rewriting Australia's Temperature History', in Jennifer Marohasy ed. *Climate Change The Facts 2020,* IPA, Melbourne.

Marohasy, J & Abbot, JW 2016, 'Southeast Australian Maximum Temperature Trends, 1887–2013: An Evidence-Based Reappraisal', *Evidence-Based Climate Science*, Elsevier, pp. 83–99.

Miles, RE 1978, 'The Origin and Meaning of Miles's Law', *Public Administration Review*, vol. 38, no. 5, pp. 399–403.

Ollier, C 2009, 'Lysenkoism and "Global Warming"', *Energy and Environment*, vol. 20, pp. 197–200.

Paltridge, G, Arking, A & Pook, M 2009, 'Trends in middle-and upper-level tropospheric humidity from NCEP reanalysis data', *Theoretical and Applied Climatology*, vol. 98, no. 3, pp. 351–359.

Pielke Jr, R 2007, *The honest broker: making sense of science in policy and politics*, Cambridge University Press, Cambridge.

Pielke Jr, R 2023, 'What the IPCC Actually Says About Extreme Weather', *The Honest Broker*, 20 July, https://rogerpielkejr.substack.com/p/what-the-ipcc-actually-says-about

Pielke Jr, R & Ritchie, J 2021, 'How climate scenarios lost touch with reality', *Issues in Science and Technology*, vol. 37, no. 4, pp. 74–83.

Pierson, P 1993, 'When Effect Becomes Cause: Policy Feedback and Political Change', *World Politics*, vol. 45, pp. 595–628.

Scafetta, N 2022, 'Advanced Testing of Low, Medium, and High ECS CMIP6 GCM Simulations Versus ERA5-T2m', *Geophysical Research Letters*, vol. 49, no. 6, e2022GL097716.

Scafetta, N 2023, 'CMIP6 GCM ensemble members versus global surface temperatures,' *Climate Dynamics*, vol. 60, no. 9–10, pp. 3091–3120.

Scafetta, N 2023a, 'CMIP6 GCM Validation Based on ECS and TCR Ranking for 21st Century Temperature Projections and Risk Assessment', *Atmosphere*, vol. 14, p. 345.

Sellitto, P, Podglajen, A, Belhadji, R, Boichu, M, Carboni, E, Cuesta, J, Duchamp, C, Kloss, C, Siddans, R, Bègue, N, Blarel, L, Jegou, F, Khaykin, S, Renard, J-B & Legras, B 2022, 'The unexpected radiative impact of the Hunga Tonga eruption of 15th January 2022', *Communications Earth & Environment*, vol. 3, no. 1, p. 288.

Suzuki, D 1997, *The Sacred Balance*, Greystone Books, Vancouver.

van Wijngaarden, WA & Happer, W 2019, 'Methane and Climate', *CO$_2$ Coalition*, 26 November, https://co2coalition.org/publications/methane-and-climate/

van Wijngaarden, WA & Happer, W 2020, 'Dependence of Earth's thermal radiation on five most abundant greenhouse gases', *arXiv preprint*, arXiv:2006.03098.

Yandle, B 1983, 'Bootleggers and Baptists – the education of a regulatory economist', *Regulation*, vol. 7, p. 12.

7: Precipitation: The Achilles Heel of Climate Models

Biasutti, M, Battisti, DS & Sarachik, ES 2003, 'The annual cycle over the Tropical Atlantic, South America, and Africa', *Journal of Climate*, vol. 16, pp. 2491–2508.

Byrne, MP & O'Gorman, PA 2018, 'Trends in continental temperature and humidity directly linked to ocean warming', *Proceedings of the National Academy of Sciences*, vol. 115, pp. 4863–4868.

Christopherson, RW & Birkeland, GH 2015, *Geosystems: Ninth Edition*, Pearson Education, Inc., Upper Saddle River, NJ, p. 195.

Christy, JR 2012, Testimony to the U.S. Senate Committee on Environment and Public Works, August 1.

Dai, A 2006, 'Precipitation characteristics in eighteen coupled climate models', *Journal of Climate*, vol. 19, pp. 4605–4630.

Dommenget, D & Rezny, M 2017, 'A caveat note on tuning in the development of coupled climate models', *Journal of Advances in Modeling Earth Systems*, vol. 10, pp. 78–97.

Done, JM, Leung, LR, Davis, CA & Kuo, B 2005, 'Understanding the value of high resolution regional climate modeling', *AMS Forum: Living with a Limited Water Supply*, paper 5.1.

Faghih, M, Brissette, FP & Sabeti, P 2022, 'Impact of correcting sub-daily climate model biases for hydrological studies', *Hydrology and Earth System Sciences*, vol. 26, 1545–1563.

Ferguglia, O, Von Hardenberg, J & Palazzi, E 2023, 'Robustness of precipitation Emergent Constraints in CMIP6 models', *Climate Dynamics*, vol. 61, pp. 1439–1450.

Golaz, J-C, Horowitz, LW & Levy, HI 2013, 'Cloud tuning in a coupled climate model: Impact on 20th century warming', *Geophysical Research Letters*, vol. 40, pp. 2246–2251.

Guan, T, Liu, Y, Sun, Z, Zhang, J, Chen, H, Wang, G, Jin, J, Bao, Z &Qi, W 2022, 'A framework to identify the uncertainty and credibility of GCMs for projected future precipitation: A case study in the Yellow River Basin, China', *Frontiers in Environmental Science*, vol. 10.

Gutowski, WJ Jr, Ullrich, PA, Hall, A, Leung LR, O'Brien, TA, Patricola, CM, Arritt, RW, Bukovsky, MS, Calvin, KV, Feng, Z, Jones, AD, Kooperman, GJ, Monier, E, Pritchard, MS, Pryor, SC, Qian, Y, Rhoades, AM, Roberts, AF, Sakaguchi, K, Urban, N & Zarzycki, C 2020, 'The ongoing need for high-resolution regional climate models – Process understanding and stakeholder information', *Bulletin of the American Meteorological Society*, vol. 101, iss. 5, E664–E683.

Hoogewind, KA, Baldwin, ME & Trapp, RJ 2017, 'The impact of climate change on hazardous convective weather in the United States: Insight from high-resolution dynamical downscaling', *Journal of Climate*, vol. 30, iss. 24, pp. 10,081–10,100.

Hostetler, SW, Bartlein, PJ, Holman, JO, Solomon, AM & Shafer, SL 2003, 'Using a regional climate model to diagnose climatological and meteorological controls of wildfire in the western United States', *Fifth Symposium on Fire and Forest Meteorology and Second International Wildland Fire Ecology and Fire Management Congress*, American Meteorological Society, Orlando, FL, paper 1.3, pp. 1–5.

Hourdin, F, Mauritsen, T, Gettelman, A, Golaz, J-C, Balaji, V, Duan, Q, Folini, D, Ji, D, Klocke, D, Qian, Y, Rauser, F, Rio, C, Tomassini, L, Watanabe, M & Williamson, D 2017, 'The art and science of climate model tuning', *Bulletin of the American Meteorological Society*, vol. 98, pp. 589–602.

IPCC 2013, Stocker, TF et al. (eds.), *Climate change 2013: The physical science basis. Contribution of Working Group I to the Fifth Assessment Report of the Intergovernmental Panel on Climate Change*, Cambridge, England, Cambridge University Press.

Kim, G, Kim, J & Cha, D-H 2022, 'Added value of high-resolution regional climate model in simulating precipitation based on the changes in kinetic energy', *Geoscience Letters*, vol. 9, iss. 38.

Kim, H, Kang, SM, Takahashi, K, Donohoe, A & Pendergrass, AG 2021, 'Mechanisms of tropical precipitation biases in climate models', *Climate Dynamics*, vol. 56, pp. 17–27.

Langford, S, Stevenson, S & Noone, D 2014, 'Analysis of low-frequency precipitation variability in CMIP5 historical simulations for southwestern North America', *Journal of Climate*, vol. 27, pp. 2735–2756.

Lee, Y-C & Wang, Y-C 2021, 'Evaluating diurnal rainfall signal performance from CMIP5 to CMIP6', *Journal of Climate*, vol. 34, pp. 7607–7623.

Legates, DR 2014, 'Climate models and their simulation of precipitation', *Energy & Environment*, vol. 25, pp. 1163–1175.

Light, CX, Arbic, BK, Martin, PE, Brodeau, L, Farrar, JT, Griffies, SM, Kirtman, BP, Laurindo, LC, Menemenlis, D, Molod, A, Nelson, AD, Nyadjro, E, O'Rourke, AK, Shriver, JF, Siqueira, L, Small, RS & Strobach, E 2022, 'Effects of grid spacing on high-frequency precipitation variance in coupled high-resolution global ocean–atmosphere models', *Climate Dynamics*, vol. 59, pp. 2887–2913.

Lindberg, C & Broccoli, AJ 1996, 'Representation of topography in spectral climate models and its effect on simulated precipitation', *Journal of Climate*, vol. 9, pp. 2641–2659.

Martinez-Villalobos, C, Neelin, JD & Pendergrass, AG 2022, 'Metrics for evaluating CMIP6 representation of daily precipitation probability distributions', *Journal of Climate*, vol. 35, pp. 5719–5743.

Soden, BJ 2000, 'The sensitivity of the tropical hydrological cycle to ENSO', *Journal of Climate*, vol. 13, pp. 538–549.

Song, H, Lin, W, Lin, Y, Wolf, AB, Donner, LJ, Del Genio, AD, Neggers, R, Endo, S & Liu, Y 2014, 'Evaluation of cloud fraction simulated by seven SCMs against the ARM observations at the SGP site', *Journal of Climate*, vol. 27, pp. 6698–6719.

Stephens, GL, L'Ecuyer, TS, Forbes, RM, Gettelmen, A, Golaz, J-C, Bodas-Salcedo, A, Suzuki, K, Gabriel, P & Haynes, J 2010, 'Dreary state of precipitation in global models', *Journal of Geophysical Research*, vol. 115.

Suzuki, K, Golaz, J-C & Stephens, GL 2013, 'Evaluating cloud tuning in a climate model with satellite observations' *Geophysical Research Letters*, vol. 40, pp. 4464–4468.

Taszarek, M, Allen, JT, Brooks, HE, Pilguj, N & Czernecki, B 2021, 'Differing trends in United States and European severe thunderstorm environments in a warming climate', *Bulletin of the American Meteorological Society*, vol. 102, iss. 2, E296–E322.

Thomassen, ED, Kendon, EJ, Sorup, HJD, Chan, S, Langen, PL, Christensen, OB & Arnbjerg-Nielsen, K 2021, 'Differences in representation of extreme precipitation events in two high resolution models', *Climate Dynamics*, vol. 57, pp. 3029–3043.

Uhe, P, Mitchell, D, Bates, PD, Allen, MR, Betts, RA, Huntingford, C, King, AD, Sanderson, BM & Shiogama, H 2021, 'Method uncertainty is essential for reliable confidence statements of precipitation projections', *Journal of Climate*, vol. 34, pp. 1227–1240.

Vicente-Serrano, SM, García-Herrera, R, Peña-Angulo, D, Tomas-Burguera, M, Domínguez-Castro, F, Noguera, I, Calvo, N, Murphy, C, Nieto, R, Gimeno, L, Gutierrez, JM, Azorin-Molina, C & El Kenawy, A 2022, 'Do CMIP models capture long-term observed annual precipitation trends?' *Climate Dynamics*, vol. 58, pp. 2825–2842.

Wang, XL, Wan, H, Zwiers, FW, Swail, VR, Compo, GP, Allan, RJ, Vose, RS, Jourdain, S & Yin, X 2011, 'Trends and low-frequency variability of storminess over western Europe 1878–2007', *Climate Dynamics*, vol. 37, pp. 2355–2371.

Wang, Y, Wang, L, Feng, J, Song, Z & Wu, Q 2023, 'A statistical description method of global sub-grid topography for numerical models', *Climate Dynamics*, vol. 60, pp. 2547–2561.

Xu, Z, Chang, A & Di Vittorio, A 2022, 'Evaluating and projecting of climate extremes using a variable-resolution global climate model (VR-CESM)', *Weather and Climate Extremes*, vol. 38, iss. 17.

Zhou, Y, Yu, R, Zhang, Y & Li, J 2023, 'Dynamic and thermodynamic processes related to precipitation diurnal cycle simulated by GRIST', *Climate Dynamics*, vol. 61, pp. 3935–3953.

Zolina, O 2014, 'Understanding hydroclimate extremes in a changing climate: Challenges and perspectives', *GEWEX News*, vol. 24, pp. 18–22.

8: Characteristics of Recent Climate Change

Caesar, L, Rahmstorf, S, Robinson, A, Fuelner G & Saba, V 2018, 'Observed fingerprint of weakening Atlantic Ocean overturning circulation', *Nature*, vol. 556, pp. 191–196.

Kalnay, E, Kanamitsu, M, Kistler, R, Collins, W, Deaven, D, Gandin, L, Iredell, M, Saha, S, White, G, Woollen, J, Zhu, Y, Chelliah, M, Ebisuzaki, W, Higgins, W, Janowiak, J, Mo, KC, Ropelewski, C, Wang, J, Leetmaa, A, Reynolds, R, Jenne, R, Joseph, D 1996, The NCEP/NCAR Reanalysis 40-year Project, *Bulletin of the American Meteorological Society*, vol. 77, pp. 437–471.

Kiehl, JT & Trenberth, KE 1997, 'Earth's annual global mean energy budget', *Bulletin of the American Meteorological Society*, vol. 78, pp. 197–208.

NOAA Physical Sciences Laboratory, Boulder Colorado, https://psl.noaa.gov/data/index.html

Riehl, H & Malkus, JS 1958, 'On the heat balance of the equatorial trough zone', *Geophysica*, vol. 6, no. 3–4, pp. 503–538.

Trenberth, KE & Caron JM 2001, 'Estimates of meridional atmosphere and ocean heat transports', *Journal of Climate*, vol. 14, pp. 3433–3437.

Trenberth, KE & Stepaniak DP 2003, 'Covariability of poleward atmospheric energy transports on seasonal and interannual timescales', *Journal of Climate*, vol. 16, pp. 3691–3705.

9: What Causes Increasing Greenhouse Gases?

Berry, E 2021, 'The impact of human CO_2 on atmospheric CO_2', *Science of Climate Change*, vol. 1, no. 2, pp. 213–249, https://doi.org/10.53234/scc202112/13

CDIAC 2017, Carbon Dioxide Information Analysis Center, ESS-DIVE Archive, https://cdiac.ess-dive.lbl.gov/

Harde, H 2017, 'Scrutinizing the carbon cycle and CO_2 residence time in the atmosphere', *Global Planetary Change*, vol. 152, pp. 19–26, http://dx.doi.org/10.1016/j.gloplacha.2017.02.009

Harde, H 2019, 'What humans contribute to atmospheric CO_2: Comparison of carbon cycle models and observations', *Earth Sciences*, vol. 8, pp. 139–158, https://doi.org/10.11648/j.earth.20190803.13

Harde, H & Salby, M 2021, 'What controls the atmospheric CO_2 level?' *Science of Climate Change*, vol. 1, no. 1, pp. 54–69, https://doi.org/10.53234/scc 202106/22

Harde, H 2023, 'Understanding Increasing Atmospheric CO_2', *Science of Climate Change*, vol. 3.1, pp. 46–67, https://doi.org/10.53234/scc202301/23

Humlum, O, Stordahl, K & Solheim, J-E 2013, 'The phase relation between atmospheric carbon dioxide and global temperature', *Global Planetary Change*, vol. 100, pp. 51–69.

IPCC, 2001, Houghton, JT et al. (eds.), *Climate Change 2001: The Scientific Basis. Contribution of Working Group I to the Third Assessment Report of the Intergovernmental Panel on Climate Change*, Cambridge, England, Cambridge University Press.

Kennedy, J, Rayner, N, Atkinson, C & Killick, R 2019, 'An ensemble data set of sea surface temperature change from 1850, The Met Office Hadley Centre HadSST.4.0.0.0 Data Set', *JGR Atmospheres*, vol. 124, pp. 7719–7763.

Koutsoyiannis, D, Onof, C, Kundzewicz, ZW & Christofides, A 2023, 'On Hens, Eggs, Temperatures and CO_2: Causal Links in Earth's Atmosphere', *Sci*, vol. 5, iss. 35, https://doi.org/10.3390/sci5030035

Levin, I, Krömer, B, Schoch-Fischer, H, Bruns, M, Münnich, M, Berdau, D, Vogel, JC & Münnich, KO 1994, 'Atmospheric $^{14}CO_2$ measurements from Vermunt, Austria, extended data up to 1983', https://cdiac.ess-dive.lbl.gov/ftp/trends/co2/vermunt.c14

Palmer, P, Eng, L, Baker, D, Chevallier, F, Bosch H & Somkuti, P 2019, 'Net carbon emissions from African biosphere dominate pan-tropical atmospheric CO_2 signal', *Nature Comm.*, https://doi.org/10.1038/s41467-019-11097-w

Salby, M & Harde, H 2021a, 'Control of atmospheric CO_2 – Part I: Relation of carbon 14 to the removal of CO_2', *Science Climate Change*, vol. 1, no. 2, pp. 177–195, https://doi.org/10.53234/scc202112/30

Salby, M & Harde, H 2021b, 'Control of Atmospheric CO_2 – Part II: Influence of Tropical Warming', *Science of Climate Change*, vol. 1, no. 2, pp. 196–212, https://doi.org/10.53234/scc202112/12

Salby, M & Harde, H 2022, 'Theory of increasing greenhouse gases', *Science of Climate Change*, vol. 2, no. 3, pp. 212–238, https://doi.org/10.53234/scc202212/17

Spencer, R, Christy, J & Braswell, D 2017, 'UAH version 6 global satellite temperature products: Methodology and results', *Asia-Pacific J. Atm. Sci.*, vol. 53, pp. 121–130.

10: Natural Climate Drivers versus Anthropogenic Drivers

Agnihotri, R, Dutta, K, Bhushan, R & Somayajulu, BLK 2002, 'Evidence for solar forcing on the Indian monsoon during the last millennium', *Earth and Planetary Science Letters*, vol. 198, pp. 521–527, https://doi.org/10.1016/S0012-821X(02)00530-7

Attolini, MR, Cecchini, S, Galli, M & Nanni, T 1987, 'The Gleissberg and 130-year periodicity in the cosmogenic isotopes in the past: The Sun as a quasi-periodic system', Proceedings of the 20th International Cosmic Ray Conference, Moscow, 4, 323, Nauka, Moscow, https://articles.adsabs.harvard.edu/pdf/1987ICRC...20d.323A

Bard, E, Raisbeck, G, Françoise, You, F & Jouzel, J 2000, 'Solar irradiance during the last 1200 years based on cosmogenic nuclides', *Tellus B*, vol. 52, pp. 985–992, https://onlinelibrary.wiley.com/doi/abs/10.1034/j.1600-0889. 2000.d01-7.x

Barrett, J, Bellamy, D & Hug, H 2006, 'On the sensitivity of the atmosphere to the doubling of the carbon dioxide concentration and on water vapour feedback', *Energy & Environment*, vol. 17, iss. 4, pp. 603–607. https://sci-hub.wf/10.1260/ 095830506778644198

Bengtsson, L & Schwartz, SE 2013, 'Determination of a lower bound on Earth's climate sensitivity', *Tellus B: Chemical and Physical Meteorology*, vol. 65, iss. 1, DOI: 10.3402/tellusb.v65i0.21533

Berry, EX 2021, 'The impact of human CO_2 of atmospheric CO_2', *Science of Climate Change*, vol. 1.2, pp. 213–249, https://scienceofclimatechange.org/wp-content/ uploads/Berry-2021-Impact-of-human-CO2.pdf

Bond, G 1997, 'A Pervasive Millennial-Scale Cycle in North Atlantic Holocene and Glacial Climates', *Science*, vol. 278, pp. 1257–1266, DOI: 10.1126/science. 278.5341.1257

CERES, 2021, The National Oceanic and Atmospheric Administration (NOAA), CERES EBAF-TOA Data: https://ceres-tool.larc.nasa.gov/ord-tool/jsp/ EBAFTOA41Selection.jsp

Chen, D, Wang, H, Sun, J & Gao, Ya 2018, 'Pacific multi-decadal oscillation modulates the effect of Arctic oscillation and El Niño southern oscillation on the East Asian winter monsoon', *International Journal of Climatology*, vol. 38, pp. 2808–2818, https://doi.org/10.1002/joc.5461

Cini Castagnoli, G, Bonino, G, Serio, M & Sonett, CP 1992, 'Common spectral features in the 5500-year record of total carbonate in sea sediments and radiocarbon in tree rings', *Radiocarbon*, vol. 34, ed. 3, pp. 798–805, DOI: https://doi.org/10.1017/S0033822200064109

Connolly, R, Soon, W, Connolly, M, Baliunas, S, Berglund, J, et al. 2021, 'How much has the Sun influenced Northern Hemisphere temperature trends? An ongoing debate', *Research in Astronomy and Astrophysics*, vol. 21, p. 31, DOI: https://doi.org/10.1017/S0033822200064109

Etminan, E, Myhre, G, Highwood, EJ & Shine, KP 2016, 'Radiative forcing of carbon dioxide, methane, and nitrous oxide: A significant revision of methane radiative forcing', *Geophysical Research Letters*, vol. 43, pp. 12614–12636, https://doi.org/10.1002/2016GL071930

Feynman, J & Fougere, PF 1984, 'Eighty-eight-year periodicity in solar-terrestrial phenomena confirmed', *Journal of Geophysical Research: Space Physics*, vol. 89, pp. 3023–3027, https://doi.org/10.1029/JA089iA05p03023

Fleming, RJ 2018, 'An updated review about carbon dioxide and climate change', *Environmental Earth Sciences*, vol. 77, 262, https://link.springer.com/article/10.1007/s12665-018-7438-y

FTPC 2022, 'The supra-long Scots pine tree-ring record for Finnish Lapland', Natural Resources Institute Finland (LUKE), viewed 6 November 2023, https://lustialab.com/data/Advance/adv7638.exz

Gats 2014, Spectral calculations tool, viewed 6 November 2023, http://www.spectralcalc.com/info/about.php

Gleissberg, W 1958, 'The eighty-year sunspot cycle', *Journal of British Astronomical Association*, vol. 68, pp. 148–152.

HadCRUT4 2021, HadCRUT4 temperature data of Met Office Hadley Centre, viewed 6 November 2023, https://www.metoffice.gov.uk/hadobs/hadcrut4/

Harde, H 2013, 'Radiation and heat transfer in the atmosphere: A comprehensive approach on a molecular basis', *International Journal of Atmospheric Sciences*, vol. 2013, ID. 503727, https://doi.org/10.1155/2013/503727

Harde, H 2017, 'Radiation transfer calculations and assessment of global warming by CO_2', *International Journal of Atmospheric Sciences*, vol. 2017, ID. 9251034, https://downloads.hindawi.com/archive/2017/9251034.pdf

Harde, H 2022, 'How Much CO_2 and the Sun Contribute to Global Warming: Comparison of Simulated Temperature Trends with Last Century Observations', *Science of Climate Change*, vol. 2.2, pp. 105–133, https://scienceofclimatechange.org/wp-content/uploads/Harde-2022-CO2-Sun-Global-Warming.pdf

HITRAN 2021, High-Resolution Transmission Molecular Absorption data base, Harvard-Smithsonian Center for Astrophysics, https://www.cfa.harvard.edu/hitran/

Hoyt, DV & Schatten, KH 1993, 'A discussion of plausible solar irradiance variations, 1700–1992', *Journal of Geophysical Research*, vol. 98, iss. A11, pp. 18895–18906, https://doi.org/10.1029/93JA01944

IPCC 2001, *Climate Change 2001, The Physical Science Basis*, TAR, (eds. Salomon S. et al.), Cambridge University Press, UK, https://www.ipcc.ch/site/assets/uploads/2018/03/WGI_TAR_full_report.pdf

IPCC 2007, *Climate Change 2007: Synthesis Report. Contribution of Working Groups I, II and III to the Fourth Assessment Report of the Intergovernmental Panel on Climate Change* [Core Writing Team, Pachauri, RK and Reisinger, A (eds.)], IPCC, Geneva, Switzerland.

IPCC 2013, *Climate Change 2011, The Physical Science Basis*, AR5, (eds. Salomon, S et al.). Cambridge University Press, UK, https://www.ipcc.ch/site/assets/uploads/2017/09/WG1AR5_Frontmatter_FINAL.pdf

IPCC 2021, *Climate Change 2021, The Physical Science Basis*, AR6, Cambridge University Press, UK, https://www.ipcc.ch/report/ar6/wg1/

Kauppinen, J, Heinonen, JT & Malmi, PJ 2014, 'Influence of relative humidity and clouds on the global mean surface temperature', *Energy & Environment*, vol. 25, iss. 2, https://doi.org/10.1260/0958-305X.25.2.389

Kerr, RA 2000, 'A North Atlantic Climate Pacemaker for the Centuries', *Science*, vol. 288, pp. 1984–1985, DOI: 10.1126/science.288.5473.1984

Kissin, YV 2015, 'A simple alternative model for the estimation of the carbon dioxide effect on the Earth's energy balance', *Energy & Environment*, vol. 26, iss. 8, 1319–1333, https://doi.org/10.1260/0958-305X.26.8.1319

Lean, J 1995, 'Construction of solar irradiance since 1610: Implications for climate change', *Geophysical Research Letters*, vol. 22, pp. 3195–3198, https://doi.org/10.1029/95GL03093

Lean, J 2004, 'Solar Irradiance Reconstruction', IGBP PAGES/World Data Center for Paleoclimatology Data Contribution Series # 2004-035, NOAA/NGDC Paleoclimatology Program.

Lean, J 2010, 'Cycles and trends in solar irradiance', *WIREs Climate Change*, vol. 1, pp. 111–122, https://doi.org/10.1002/wcc.18

Lewis, N & Curry JA 2015, 'The implications for climate sensitivity of AR5 forcing and heat uptake estimates', *Climate Dynamics*, vol. 45, pp. 1009–1023, https://doi.org/10.1007/s00382-014-2342-y

Lin, YC, Fan, CY, Damon, PE & Wallick, EI 1975, 'Long-term modulation of cosmic-ray intensity and solar activity cycles', 14th International Cosmic Ray Conference, Germany, Munchen, vol. 3, pp. 995–999. Max-Planck-Institut für extraterrestrische Physik, Germany, https://adsabs.harvard.edu/full/1975ICRC....3..995L

Meinshausen, M, Nicholls, MRJ, Lewis, J, Gidden, MJ, Vogel, E, et al. 2020, 'The shared socio-economic pathway (SSP) greenhouse gas concentrations and their extensions to 2500', *Geoscientific Model Development*, vol. 13, pp. 3571–3605, https://doi.org/10.5194/gmd-13-3571-2020

Miskolczi, FM & Mlynczak, MG 2004, 'The greenhouse effect and the spectral decomposition of the clear-sky terrestrial radiation', *Idöjaras*. Vol. 108, pp. 209–251, http://owww.met.hu/idojaras/IDOJARAS_vol108_No4_01.pdf

Myhre, G, Highwood, EJ, Shine, KP & Stordal, F 1998, 'New estimates of radiative forcing due to well-mixed greenhouse gases', *Geophysical Research Letters*, vol. 25, pp. 2715–2718, https://doi.org/10.1029/98GL01908

NOAA 2021, NCEP/NCAR Reanalysis Data, https://www.esrl.noaa.gov/psd/cgi-bin/data/timeseries/timeseries1.pl

Ohmura, A 2001, 'Physical basis for the temperature-based melt-index method', *Journal of Applied Meteorology and Climatology*, vol. 40, pp. 753–761, https://doi.org/10.1175/1520-0450(2001)040<0753:PBFTTB>2.0.CO;2

Ollila, A 2012, 'The roles of greenhouse gases in global warming', *Energy & Environment*, vol. 23, iss. 5, pp. 781–799, https://doi.org/10.1260/0958-305X.23.5.781

Ollila, A 2013, 'Dynamics between clear, cloudy and all-sky conditions: cloud forcing effects', *Journal of Chemical Biological Physical Sciences*, vol. 4, iss. 1, pp. 557–575, https://www.researchgate.net/publication/274958251_Dynamics_between_clear_cloudy_and_all-sky_conditions_Cloud_forcing_effects

Ollila, A 2017, 'Semi empirical model of global warming including cosmic forces, greenhouse gases, and volcanic eruptions', *Physical Science International Journal*, vol. 15, iss. 2, pp. 1–14, DOI:10.9734/PSIJ/2017/34187

Ollila, A 2020, 'Analysis of the simulation results of three carbon dioxide (CO_2) cycle models', *Physical Science International Journal*, vol. 23, iss. 4, pp. 1–19, DOI: 10.9734/PSIJ/2019/v23i430168

Ollila, A 2021, 'Global Circulations Models (GCMs) simulate the current temperature only if the shortwave radiation anomaly of 2000s has been omitted', *Physical Science International Journal*, vol. 40, iss. 17, pp. 45–52, DOI:10.9734/CJAST/2021/v40i1731433

Ollila, A 2023a, 'Natural climate drivers dominate in the current warming', *Science of Climate Change*, vol. 3.3, pp. 290–326, https://scienceofclimatechange.org/wp-content/uploads/Ollila-2023-Global-Warming-Review.pdf

Ollila, A 2023b, 'Radiative forcing and climate sensitivity of carbon dioxide (CO2) fine-tuned with CERES data', *Current Journal of Applied Science and Technology*, vol. 42, iss. 46, pp. 111-133, DOI: 10.9734/CJAST/2023/v42i464300

Ollila, A & Timonen, M 2023, 'Two main temperature periodicities related to planetary and solar activity oscillations', HAL Open Science, viewed 6 November 2023, https://hal.science/hal-04160543

Otto, A, Otto, FEL, Boucher, O, Church, J, Hegerl, G, et al. 2013, 'Energy budget constraints on climate response', *Nature Geoscience*, vol. 6, iss. 6, pp. 415–416, https://doi.org/10.1038/ngeo1836

Patterson, RT, Prokoph, A & Changa, A 2004, 'Late Holocene sedimentary response to solar and cosmic ray activity influenced climate variability in the NE Pacific', *Sedimentary Geology*, vol. 172, pp. 67–84, https://doi.org/10.1016/j.sedgeo.2004.07.007

Peristykh, AN & Damon, PE 2003, 'Persistence of the Gleissberg 88-year solar cycle over the last ~ 12,000 years: Evidence from cosmogenic isotopes', *Journal of Geophysical Research: Space Physics*, vol. 108, iss. A1, p. 1003, https://doi.org/10.1029/2002JA009390

Ramanathan, V, Cicerone, R, Singh, H & Kiehl, I 1985, 'Trace gas trends and their potential role in climate change', *Journal of Geophysical Research*, vol. 90, pp. 5547–5566, https://doi.org/10.1029/JD090iD03p05547

Scafetta, N 2010, 'Empirical evidence for a celestial origin of the climate oscillations and its implications', *Journal of Atmospheric and Solar-Terrestrial Physics*, vol. 72, pp. 951–970, https://doi.org/10.1016/j.jastp.2010.04.015

Schlesinger, ME & Ramankutty, N 1994, 'An oscillation in the global climate system of period 65–70 years', *Nature*, vol. 367, 723–726, https://doi.org/10.1038/367723a0

Schildknecht, D 2020, 'Saturation of the infrared absorption by carbon dioxide in the atmosphere', *International Journal of Modern Physics B*, https://arxiv.org/pdf/2004.00708.pdf

Schmidt, GA, Ruedy, R, Miller, RL & Lacis, AA 2010, 'Attribution of the present-day total greenhouse effect', *Journal of Geophysical Research*, vol. 115, iss. D20, D20106, https://agupubs.onlinelibrary.wiley.com/doi/full/10.1029/2010JD014287

Schwabe, SH 1843, 'Sonnenbeobachtungen im Jahre 1843 (in German)', Observations of the Sun in the year 1843, *Astronomische Nachrichten*, vol. 21, pp. 233–236, 10.1002/asna.18440211505

Smith, CJ, Kramer, RJ, Myhre, G, Forster, PM & Soden, BJ, et al. 2018, 'Understanding rapid adjustments to diverse forcing agents', *Geophysical Research Letters*, vol. 45, iss. 21, pp. 2023–2031, https://doi.org/10.1029/2018GL079826

UAH 2022, UAH MSU temperature data set of lower troposphere, viewed 6 November 2023, http://vortex.nsstc.uah.edu/data/msu/v6.0beta/tlt/uahncdc_lt_6.0beta5.txt

Velasco Herrera, VM, Mendoza, B &Velasco Herrera, G 2015, 'Reconstruction and prediction of the total solar irradiance: From the Medieval Warm Period to the 21st century', *New Astron*, vol. 34, pp. 221–233, https://sci-hub.wf/10.1016/j.newast.2014.07.009

Vinther, BM, Jones, PD, Briffa, KR, Clausen, HB, Andersen, KK, Dahl-Jensen, D & Johnsen SJ 2010, 'Climatic signals in multiple highly resolved stable isotope records from Greenland', *Quaternary Science Reviews*, vol. 29, pp. 522–538, https://doi.org/10.1016/j.quascirev.2009.11.002

Vinther, BM 2011, 'The medieval climate anomaly in Greenland ice core data', *PAGES news*, vol. 19, iss. 1, p. 27, https://pastglobalchanges.org/sites/default/files/2022-04/Vinther_2011_1_27.pdf

Wijngaarden, W & Happer, W 2020, 'Dependence of Earth's thermal radiation on five most abundant greenhouse gases', arXiv, https://arxiv.org/abs/2006.03098

11: Atmospheric CO$_2$ – Natural and Anthropogenic Contributions of the Past Half-century

Alexander, MA, Halimeda, K & Nye, JA 2014, 'Climate variability during warm and cold phases of the Atlantic Multidecadal Oscillation (AMO) 1871–2008', *Journal of Maritime Systems,* vol. 133, pp. 14–26.

Alheit, J, Licandro, P, Coombs, S, Garcia, A, Giráldez, A, Santamaría, MTG, Slotte, A & Tsikliras, AC 2014, 'Atlantic Multi-decadal Oscillation (AMO) modulates dynamics of small pelagic fishes and ecosystem regime shifts in the eastern North and Central Atlantic', *Journal of Maritime Systems,* vol. 131, pp. 21–35.

Earth System Science Center 2022, Global Temperature Report, Monthly Average Data. 2022, viewed 21 June 2023, https://www.nsstc. uah.edu/data/msu/v6.0/tlt/uahncdc_lt_6.0.txt

Feely, RA, Sabine, CL, Millero, FJ, Dickson, AG, Fine, RA, Carlson, CA, Toole, J, Joyce, TM, Smethie, WM, McNichol, AP & Key, RM 2008, 'Carbon dioxide, hydrographic, and chemical data obtained during the R/V *Knorr* repeat hydrography cruise in the North Atlantic Ocean: CLIVAR CO_2 Sections A20_2003 (22 September–20 October, 2003) and A22_2003 (23 October– 13 November, 2003)', ed. A. Kozyr, ORNL/CDIAC-154, NDP-089, Carbon Dioxide Information Analysis Center, Oak Ridge National Laboratory, U.S. Department of Energy, Oak Ridge, Tennessee.

Goericke, R & Fry, B 1994, 'Variations of marine plankton $\delta^{13}C$ with latitude, temperature, and dissolved CO_2 in the world ocean', *Global Biogeochemical Cycles*, vol. 8, pp. 85–90.

Gruber, N, Bakker, DCE, DeVries, T, Gregor, L, Hauck, J, Landschützer, P, McKinley, GA & Muller, JD 2023, 'Trends and variability in the ocean carbon sink', *Nature Reviews Earth and Environment*, vol. 4, pp. 119–134.

Hare, SR & Mantua, NJ 2000, 'Empirical evidence for North Pacific regime shifts in 1977 and 1989', *Progress in Oceanography*, vol. 47, pp. 103–145.

Hoffman, DW & Rasmussen, C 2022, 'Absolute carbon stable isotope ratio in the Vienna Peedee Belemnite isotope reference determined by 1H NMR spectroscopy', *Analytical Chemistry*, vol. 94, pp. 5240–5247.

Holzer, M & DeVries, T 2022, 'Source-labeled anthropogenic carbon reveals a large shift of preindustrial carbon from the ocean to the atmosphere', *Global Biogeochemical Cycles*, vol. 36, iss. 10.

Keeling, CD, Piper, SC, Bacastow, RB, Wahlen, M, Whorf, TP, Heimann, M & Meijer, HA 2001, 'Exchanges of Atmospheric CO_2 and $13CO_2$ with the Terrestrial Biosphere and Oceans from 1978 to 2000', in *Global Aspects*; SIO Reference Series, no. 01–06, Scripps Institution of Oceanography: San Diego, CA, USA, 2001, viewed 21 June 2023, https://scrippsco2.ucsd.edu/data/atmospheric_ co2/sampling_stations.html

Kroopnick, PM 1985, 'The distribution of ^{13}C of ΣCO_2 in the world oceans', *Deep-Sea Research*, vol. 32, pp. 57–84.

NOAA n.d., Nino 3.4 SST Index, viewed 15 April 2023, https://psl.noaa.gov/gcos_wgsp/Timeseries/Data/nino34.long.data

NOAA, The global conveyor belt, viewed 21 June 2023, https://oceanservice.noaa.gov/education/tutorial_currents/05conveyor2.html

Ohmoto, H 1986, 'Stable isotope geochemistry of ore deposits', *Reviews in Mineralogy*, vol. 16, pp. 491–559.

O'Leary, M 1988, 'Carbon isotopes in photosynthesis fractionation techniques may reveal new aspects of carbon dynamics in plants', *BioScience*, vol. 38, pp. 328–336.

Quirk, T 2012, 'Did the global temperature trend change at the end of the 1990s?' *Asia-Pacific Journal of Atmospheric Sciences*, vol. 48, pp. 339–344.

Quirk, T & Asten M 2023, 'Atmospheric CO_2 Isotopic Variations, with Estimation of Ocean and Plant Source Contributions', *Atmosphere*, vol 14, p. 1623.

12: Warming by Carbon Dioxide is Too Small to Matter, According to Will Happer

Barrett, J 2005, 'Greenhouse molecules, their spectra and function in the atmosphere', *Energy and Environment*, vol. 16. no. 6, pp. 1037–1045.

Idso, C 2017, 'Carbon dioxide and plant growth', in J Marohasy (ed), *Climate Change the Facts 2017*, Institute of Public Affairs, Australia, pp. 189–200.

Happer, W 2023, 'CO_2, The Gas of Life', A talk for the Institute of Public Affairs – IPA The Voice for Freedom, September, https://www.youtube.com/watch?v=v2nhssPW77I

Marohasy, J 2020, 'Rewriting Australia's Temperature History', in J Marohasy (ed), *Climate Change The Facts 2020*, Institute of Public Affairs, Australia, pp. 241–252.

Philander, G 1998, *Is the Temperature Rising? The Uncertain Science of Global Warming*, Princeton University Press, p. 262.

Thunberg, G 2022, *The Climate Book*, Penguin Random House, p. 446.

van Wijngaarden, WA & Happer, W 2020, Dependence of Earth's Thermal Radiation on Five Most Abundant Greenhouse Gases, Fig. 4, viewed 23 October 2023, https://arxiv.org/pdf/2006.03098

13: Volcanoes and Climate

Evan, S, Brioude, J, Rosenlof, KH, Gao, R, Portmann, RW, Zhu, Y, Volkamer, R, Lee, CF, Metzger, JM, Lamy, K, Walter, P, Alvarez, SL, Flynn, JH, Asher, E, Todt, M, Davis, SM, Thornberry, T, Vömel, H, Wienhold, FG, Stauffer, RM, Millán, L, Santee, ML, Froidevaux, L & Read, WG 2023, 'Rapid ozone depletion after humidification of the stratosphere by the Hunga Tonga Eruption', *Science*, vol. 382, iss. 6668.

Guo, S, Bluth, GJS, Rose, WI, Watson, IM & Prata, AJ 2004, 'Re-evaluation of SO_2 release of the 15 June 1991 Pinatubo eruption using ultraviolet and infrared satellite sensors', *Geochem. Geophys. Geosyst.*, vol. 5, Q04001.

Hofmann, DJ, Oltmans, SJ, Komhyr, WD, Harris, JM, Lathrop, A, Langford, AO, Deshler, T, Johnson, BJ, Torres, A & Matthews, WA 1994, 'Ozone loss in the lower stratosphere over the United States in 1992-1993: Evidence for heterogeneous chemistry of the Pinatubo aerosol', *Geophysical Research Letters*, vol. 21, pp. 65–68.

Jenkins, S, Smith, C, Allen, M & Grainger, R 2023, 'Tonga eruption increases chance of temporary surface temperature anomaly above 1.5 °C', *Nature Climate Change,* vol. 13, pp. 127–129.

McCormick, MP & Swissler, TJ 1983, 'Stratospheric aerosol mass and latitudinal distribution of the El Chichón eruption cloud for October 1982', *Geophysical Research Letters*, vol. 10, pp. 877–880.

Raible, CC, Brönnimann, S, Auchmann, R, Brohan, P, Frölicher, TL, Graf, HF, Jones, P, Luterbacher, J, Muthers, S, Neukom, R, Robock, A, Self, S, Sudrajat, A, Timmreck, C & Wegmann, M 2016, 'Tambora 1815 as a test case for high impact volcanic eruptions. Earth system effects', *WIREs Climate Change*, vol. 7, no. 4, pp. 569–589.

Sigl, M, Winstrup, M, McConnell, JR, Welten, KC, Plunkett, G, Ludlow, F, Büntgen, U, Caffee, M, Chellman, N, Dahl-Jensen, D, Fischer, H, Kipfstuhl, S, Kostick, C, Maselli, OJ, Mekhaldi, F, Mulvaney, R, Muscheler, R, Pasteris, DR, Pilcher, JR, Salzer, M, Schüpbach, S, Steffensen, JP, Vinther BM & Woodruff, TE 2015, 'Timing and climate forcing of volcanic eruptions for the past 2,500 years', *Nature*, vol. 523, pp. 543–549.

14: Simulation of Winter Warming after Volcanic Eruptions

Bittner, M, Schmidt, H, Timmreck, C & Sienz, F 2016, 'Using a large ensemble of simulations to assess the Northern Hemisphere stratospheric dynamical response to tropical volcanic eruptions and its uncertainty', *Geophysical Research Letters*, vol. 43, pp. 9324–9332.

Charlton-Perez, A, et al. 2013, 'On the lack of stratospheric dynamical variability in low-top versions of the CMIP5 models', *Journal of Geophysical Research Atmosphere*, vol. 118, pp. 2494–2505.

Chen, X, et al. 2020, 'Boreal Winter Surface Air Temperature Responses to Large Tropical Volcanic Eruptions in CMIP5 Models', *Journal of Climate*, vol. 33, pp. 2407–2426.

Chylek, P, Folland, CK, Klett, JD, Wang, M, Lesins, G & Dubey, MK 2022, 'Annual mean Arctic Amplification 1970–2020: Observed and simulated by CMIP6 climate models', *Geophysical Research Letters*, vol. 49, e2022GL099371.

Chylek, P, Folland, CK, Klett, JD, Wang, M, Lesins, G & Dubey, MK 2023a, 'High Values of the Arctic Amplification in the Early Decades of the 21st Century: Causes of Discrepancy by CMIP6 Models Between Observation and Simulation', *Journal of Geophysical Research Atmosphere*, vol. 128, e2023JD039269.

Chylek, P, Folland, CK, Klett, JD, Wang, M, Lesins, G & Dubey, MK 2023b, 'Arctic Amplification in the Community Earth System Models (CESM1 and CESM2)', *Atmosphere*, vol. 14, p. 820.

Driscoll, S, Bozzo, A, Gray, L, Robock, A & Stenchikov, G 2012, 'Coupled Model Intercomparison Project 5 (CMIP5) simulations of climate following volcanic eruptions', *Journal of Geophysical Research Atmosphere*, vol. 117, D17105.

Hausfather, Z, Marvel, K, Schmidt, G, Nielsen-Gammon, J & Zelinka, M 2022, 'Climate simulations: recognize the "hot model" problem', *Nature,* vol. 605, pp. 26–29.

Irvine, P, Emanuel, K, He, I, Horowitz, L, Vecchi, G & Keith, D 2019, 'Halving warming with idealized solar geoengineering moderates key climate hazards', *Nature Climate Change*, vol. 9, pp. 295–299.

MacMartin, D, Kravitz, B, Long, J & Rasch, P 2016, 'Geoengineering with stratospheric aerosols: What do we not know after a decade of research?', *Earth's Future*, vol. 4, pp. 543–554.

Rashid, H, Sullivan, A, Dix, M, Daohua, B, Mackallah, C, Ziehn, T, Dobrohotoff, P, O'Farrell, S, Harman, I, Bodman, R & Marsland, S 2022, 'Evaluation of climate variability and change in ACCESS historical simulations for CMIP6', *Journal of Southern Hemisphere Earth Systems Science*, vol. 72, ed. 2, pp. 73–92.

Robock, A & Mao, J 1992, 'Winter warming from large volcanic eruptions', *Geophysical Research Letters*, vol. 19, pp. 2405–2408.

Robock, A 2000, 'Volcanic eruptions and climate', *Reviews of Geophysics*, vol. 38, pp. 191–219.

Semmler, T, Danilov, S, Gierz, P, Goessling, HF, Hegewald, J, Hinrichs, C, Koldunov, N, Khosravi, N, Mu, L, Rackow, T, Sein, D, Sidorenko, D, Wang, Q & Jung, T 2020, 'Simulations for CMIP6 with the AWI climate model AWI-CM-1-1', *Journal of Advances in Modeling Earth Systems*, vol. 12, iss. 9.

Stenchikov, G, Hamilton, K, Stouffer, RJ, Robock, A, Ramaswamy, V, Santer, B & Graf, H-F 2006, 'Arctic Oscillation response to volcanic eruptions in the IPCC AR4 climate models', *Journal of Geophysical Research*, vol. 111.

Zambri, B & Robock, A 2016, 'Winter warming and summer monsoon reduction after volcanic eruptions in Coupled Model Intercomparison Project 5 (CMIP5) simulations', *Geophysical Research Letters*, vol. 43, pp. 10920–10928.

Ziehn, T, Chamberlain, M, Law, R, Lenton, A, Bodman, R, Dix, M, Stevens, L, Wang, Y-P & Srbinovsky, J 2020, 'The Australian Earth System Model: ACCESS-ESM1.5', *Journal of Southern Hemisphere Earth Systems Science*, vol. 70, pp. 193–214.

Thanks to Supporters of
Climate Change: The Facts 2025

Climate Change: The Facts 2025 is a publication of the Institute of Public Affairs (IPA), an independent research and education organisation dedicated to securing, preserving and strengthening the foundations of economic and political freedom in Australia for the next generation.

The IPA neither seeks nor accepts government funding, and so is able to review the outputs of government-funded science institutions from a purely objective standpoint, and can also draw the implications for climate policy without being concerned as to whether these contradict official pronouncements of governments or national or international climate bodies.

The IPA can maintain this independence as it relies for funding entirely on voluntary contributions from more than 9,100 members, overwhelmingly in Australia but with a number of supporters scattered across the world. Anyone can apply to be a member of the IPA, at:

www.ipa.org.au/join

In particular, *Climate Change: The Facts 2025* was made possible by the 676 people in Australia and across the world who donated to our special fundraising campaign.

In the following pages we give special recognition to the 237 people who gave consent for their names to appear as having supported the research in this volume. The IPA expresses its deep appreciation for the contributions made and this support. Many of these people have supported one or more of the previous books in the series.

As an IPA member you will receive regular updates on our research, which also appears at **www.ipa.org.au**. Further information relevant to the volume you have in your hand can be found at:

www.climatechangethefacts.org.au.

IPA

This book made possible through the generous contributions of:

ANDREW ABERCROMBIE, GERARD ABRAMS, TIM ABRAMS, ANTHONY ADAIR, STEPHEN AINSWORTH, IAN AIREY, DAVID WESTON ALLEN, HONG AN, RICHARD ANDERSEN, CAM ANDERSON, KENNETH S ANDERSON, GRAHAM ANGELL, STUART ASHTON, GARY BACON, CHARLES BARNES, K & M BARNES, ESTIE BAV, STEVEN BEIKOFF, DIANA BENNETT, GEORGE BINDLEY, DON BISHOP, ROBERT BLACKMORE, TOM BOSTOCK, JOHN RAYMOND REIS BOTTOMS, GRAHAM BRADLEY, DAVID & HALINA BRETT, DAVID BROOKES, GEOFFREY BROWN, ANDREW BROWNE, GREG BUCHANAN, EDWIN & FAY BURGE, ANDREW BUTTFIELD, LEN & WENDY CARLSON, MYRON CAUSE, PETER CESCO, JAMES SCOTT CHALMERS, JOHN CHAMBERS, PETER CHAMPNESS, G S CHOMLEY, EDEK CHOROS, MICHAEL CLANCY, CARMEL CLARKE, BILL CLOUGH, DAVID COATH, RICHARD COLEBATCH, PHILLIP COLLARD, BRIAN COMBLEY, JOHN COMPTON, LES COOPER, RICHARD CORBETT, LOUISE M M CORDNER, JOHN & JULIA CORDUKES, ROBERT COSTA, GEOFFREY COTTRELL, DAVID COURT, FRANZIE CUMMINGS, BARRY CUSACK, PETER DARVALL, CHRISTOPHER DE GUINGAND, RUSSELL CHARLES DE LA LANDE, GERRIT & BARBARA DE NYS, ANN DE VOS, JOHN DIXON, LYLE DODWELL, JOHN DOHERTY, CRAIG DONOHUE, PENELOPE DRAKE, AERT DRIESSEN, JOHN & CAROL DUDGEON, BRUCE DUFFELL, JAMES WALTER SYDNEY DURRANT, ALEX EGAN, JOHN & LIZ EGAN, MALCOLM EGLINTON, MIKE ELLIOTT, PHILLIP ELLIOTT, RICHARD EVANS, JIM EXCELL, ANTHONY FAIRBAIRN, NEIL FEARIS, JOHN FENWICK, BELINDA FISHER, TIM FITZPATRICK, JIM FLETCHER, CHRIS FLIPO, NICHOLAS J FORD, JOHN FOWLER, PETER FRANKLIN, JEFF FRASER, ROSS A FRICK, BRUCE & PETA FRYER, TONY GALL, DANNY & MARIAN GEORGESON, ROGER GIBBONS, CHRISTOPHER GOLIS, RUPERT GOOD, RODERICK GRANT, DEBRA GRAVES, MICHAEL GRAY, MICHAEL GREEN, JAMES GRIBBIN, PETER W GRIFFIN, ANDRZEJ GURBA, RODNEY HACKETT, GARY HAMMOND, KEN HANDLEY, COLIN HARKNESS, GARTH HARRIS, RICHARD HARRISON, DEAN HARVEY, GRAEME HAUSSMANN, BRECK HAYWARD, AMANDA HEMETSBERGER, GEOFF HEMM, WARWICK HERBERT, ROBERT HERCUS, ANDREW & REBECCA HEWETT, JOHN HILL,

This book made possible through the generous contributions of:

GEOFF HONE, JOHN HULL, JOHN ILLINGWORTH, GRAEME INKSTER, ALAN IRVING, EMMA JEFFREY, LYN JEFFRIES, DAVID BRUCE JOHNSTON, GERARD A JOSEPH, JOHN KEEBLE, ALEXANDER KISS, PETER KNIGHT, GLEN LAMPERD, BILL LAVER, RON LESH, STEPHEN LIPPLE, PHILLIP LOCKYER, IAN LOVEGROVE, DAVID LUBOWSKI, JOHN MACDOUGALL, WARWICK J MACKAY, PETER MANGER, ROBERT MASTERMAN, MARK MCCAULEY, STEPHEN MCCREADY, GREG & SHARON MCDERMANT, KEN MCDONALD, STUART MCGILL, MALCOLM MCKELLAR, TONY MCLEAN, WARREN MCLEAN, STEPHEN MCPHERSON, LACHLAN MCTAGGART, PAUL MICHAEL, NOELINE M MITCHELL, GORDON MOCK, MAURICE MOLAN, JAMIE MONTGOMERY, DAMON MOON, PATRICK MOORE & KATHIE DERHAM MOORE, MIKE MOYLAN, KEIRAN MULHALL, MARION MURPHY, EDWARD NEALON, ALEXANDRA NEELS, ROBERT S NIXON, BOB OFFICER, PAUL O'KEEFFE, COLIN & HELEN O'NEIL, KNOX O'NEIL, BERNARD O'SHEA, TERRY PAYTON, MICHAEL PECK, JONATHON PERKS, ANDREW PETHEBRIDGE, BRUCE PHILLIPS, DAVID PHILLIPS, WES PICKERING, DAVID PIERCE, RONALD PITCHER, TREVOR PROWSE, PATRICK PURCELL, MARK RADKE, RICHARD RANGER, WARWICK READING, KATE & MIKE RIBOT DE BRESSAC, BRIAN ROBSON, VICTOR RUDEWYCH, ROBERT RUSSELL, MURRAY SANDLAND, ANTHONY SCHNEIDER, GUDRUN SCHUCK, MAY SCOTT, JOHN & JOAN SHRAPNEL, MARIO DAVID SIMIC, JIM SIMPSON, MARIAN SKWARNECKI, TONY SMITH, ROSS SMYTH-KIRK, LAUREN SOMMERFELD, DAVID SPEEDY, GUIDO STALTARI, KERRY STEGGLES, MICHAEL STEYN, JOHN STONE, CHRIS STORY, J F STYRING, DAVID SUTHERLAND, JOHN SWARBRICK, STUART TAIT, PETER TRACEY, IAN TRISTRAM, PETER TULLOCH, EWEN TYLER, STEPHEN VOOGT, PHILIP WALKER, JIM WALLACE, MARK J WALLAND, IAN WALSH, MICHAEL WALSH, INGVAR WARNHOLTZ, TIM & ALI WATSON, DAVID WATT, BOB WHITE, ALEXANDER WHITE, GORDON WHITE, TERENCE WHITFIELD, TREVOR WIGHT, KYLE WIGHTMAN, PETER G WILLIAMSON, ANDREW WILSON, ADRIAN WISCHER, JOHN WYLD, RANDALL & ROSEMARY WYNN, RAYMOND YEOW, SOPHIE YORK, BILL YOUNG, PAULINE YOUNG, NORMAN ZILLMAN